THE HISTORIC

# BARNS

OF

# SOUTHEASTERN PENNSYLVANIA

## Architecture & Preservation, Built 1750–1900

GREGORY D. HUBER

Schiffer Publishing Ltd

4880 Lower Valley Road • Atglen, PA 19310

# Dedication

To John Huber Jr.—first and foremost—my dad,
whose magnificent sense of wonderment shall remain with me always

To Eric Sloane—the man who changed my life

"To those who can hear it, the song sung
by barns and wooden structures is precious music."
—Eric Sloane in *American Barns and Covered Bridges*

Other Schiffer Books on Related Subjects:

*American Barns*, Stanley Schuler, ISBN 978-0-88740-145-9

*Barns of Cape Cod*, Joan Dillon, ISBN 978-0-7643-2564-9

*Living Barns: How to Find and Restore a Barn of Your Own*, Ernest Burden, ISBN 978-0-7643-2410-9



Second printing 2018

Library of Congress Control Number: 2017938648



Type set in Trade Gothic & A Caslon

ISBN: 978-0-7643-5319-2
Printed in China

Published by Schiffer Publishing, Ltd.
4880 Lower Valley Road
Atglen, PA 19310
Phone: (610) 593-1777; Fax: (610) 593-2002
E-mail: Info@schifferbooks.com
Web: www.schifferbooks.com

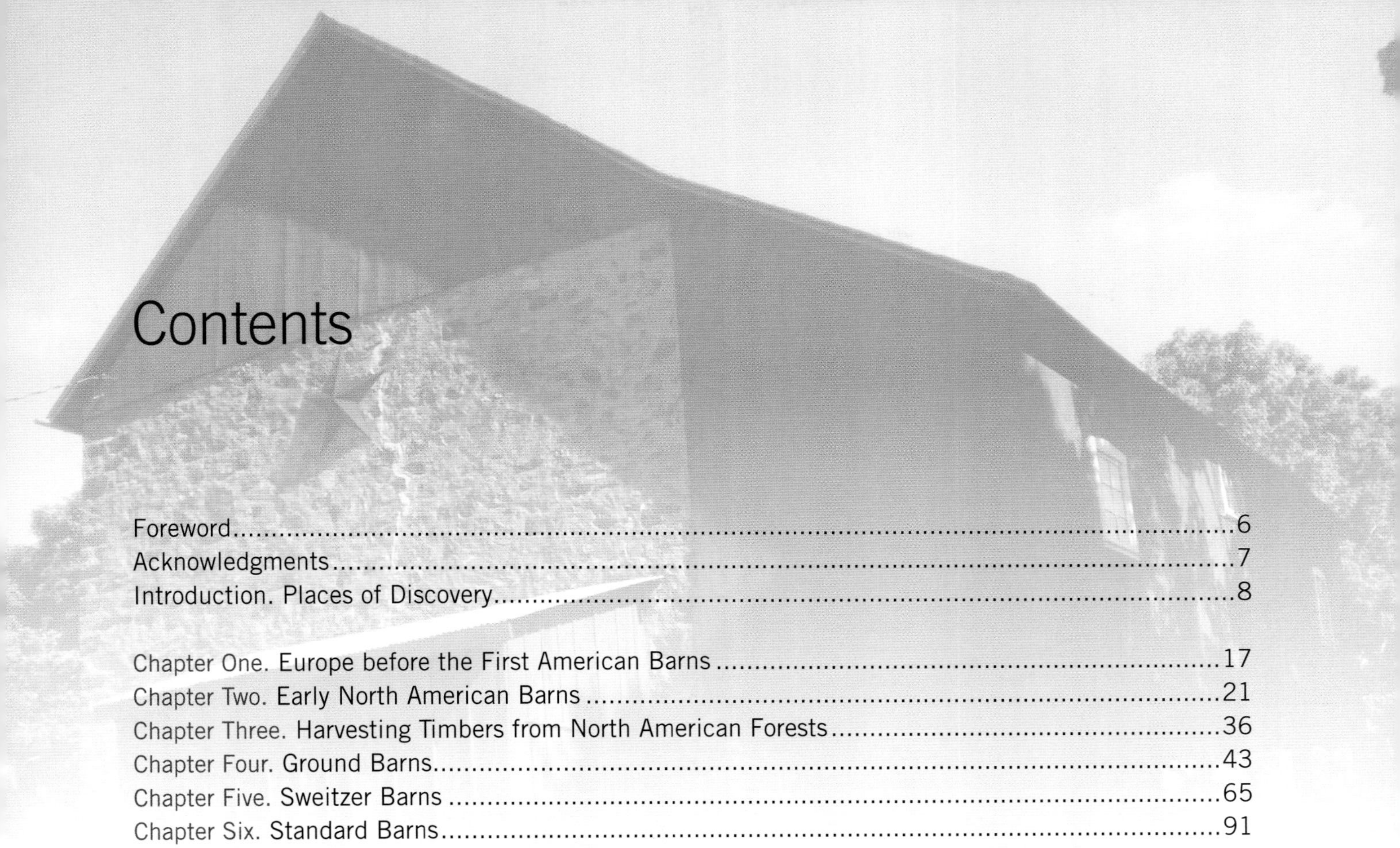

# Contents

# Foreword

Southeastern Pennsylvania's rural landscape is dominated by barns, and architectural historian Greg Huber has named this region the state's hearth. This fertile ground was settled during the seventeenth and eighteenth centuries by national and ethnic groups from Western Europe who brought their architectural traditions with them. They established farming practices and architectural styles that produced a varied agricultural landscape. The farming activity required the construction of vernacular houses and distinctive barns. The larger utilitarian barn thus became the most dominant structure and a defining element of this landscape.

In this book, Huber examines in scholarly detail the story of these barns and their development over the course of 200 years. His prodigious effort, extending over a period of forty years, required the examination of thousands of barns, many in great detail. This may never again be attempted, let alone accomplished. The resulting volume examines all the barn forms and many styles and types throughout the hearth area. A classification system brings order to a complex range of barn structures. Detailed explanations cover all phases of barn design, function, and construction. More than 250 photographs document, illustrate, and explain the complexities involved in conceiving, building, using, and maintaining vernacular barns in this important region. This definitive book has now been written. Its record will prevail over time, even though many of the barns may not.

—Robert Ensminger, author of *The Pennsylvania Barn: Its Origin, Evolution, and Distribution in North America* (Johns Hopkins University Press, 1992).

# Acknowledgments

I am deeply grateful to my father, John Huber Jr. Life in my teens and beyond was a multifaceted challenge, and if it had not been for my dad, God only knows where I would have ended up. He grew to know exactly what barns meant to me. At one point, he sent a letter to a friend proclaiming, "We have seen every barn within fifty miles of our house." Admittedly, that was a fabrication, but only in a purely physical sense. Dad and I saw far more barns than that by the power of the vision he imparted to me. He is gone, but I know he has seen all phases of this book from start to finish.

If not for Eric Sloane, I may never have started looking at barns. He could render barns like no other artist. His gravestone in Kent, Connecticut, reads: "God only knows I tried." His thirty-three books on Americana prove his statement. He, like my dad, was an amazing man. I am sorry he never got to see this book.

Perhaps it is strange to mention a four-legged animal, but Jackie, my wonderful German shepherd, was a powerful and pervasive presence in my life for almost fifteen years. She accompanied me to more barns than any human ever did, and she was the best friend I ever had. I miss her enormously.

Nancy Zeigler, who hails from Camp Hill, west of Harrisburg, has a magnificent eye for so many things, including the architectural. She was my travel companion on trips to see barns, including some extremely old Dutch-American barns in central New Jersey.

Robert Ensminger and I have spoken hundreds of times on the topic of barns, and we have gone on dozens of barn trips. I will always cherish my friendship with Bob.

Thank you to other friends in the barn world: Peter Sinclair, Claire K. Tholl (bless her heart), James Weaver, George Turrell, Chris Albright, Tim Adriance, John Adriance, Abbott Lowell Cummings, Jack Sobon, Lee and Carolina Lazzari, Doctor and Mrs. Lord, Alf Evers, Freddie Charles, Cecil Hewitt, J. T. Smith, Richard Babcock, Clarke Blair, Sheila Miller, Ken Sandri, Alan Keyser, Robert Bucher, Laura Walker, Jeffrey Marshall, Benuel Glick, Melvin Landis, Jim and Kathy Coker, John Heyl, Chris Witmer, Joe Glass, Henry Glassie, John Gockley, Alice Gerard, Marian Moffett, Patrick Donmoyer, Priscilla DeLeon, Rose Parry, Carla Cielo, Judy Lengle, Gail Hunton, Alex Greenwood, Elric Endersby, John Kuhns, Johnny Claypoole, Eric Claypoole (he watches that compass), Charles Peterson, Gerry Milnes, Willis "Skip" Barshied, Shirley Dunn, Vincent Schaeffer, John Stevens, Syd Silverman (God bless this generous man), Thomas McCullough of the Moravian Archives in Bethlehem, and the owners of the Sheffey Ground barn near Limerick, Pennsylvania, both past and present. I greatly regret never meeting Alfred Shoemaker.

Thank you to The Athenaeum of Philadelphia, which awarded me a Charles Peterson Fellowship in 2002 and extended their generosity after the initial grant. I particularly want to thank Roger Moss and Bruce Laverty. Their two grants helped me to do an extensive study on early German roof structures.

To my brother, Robert Huber, who awakened my awareness to barns when I was very young in the mid-1950s. Rob had a favorite red barn on Western Highway in West Nyack, New York, near where my dad grew up.

To my brother, Barry Huber, who went to a sports memorabilia show in the mid-1990s wearing a "Save the Tice Barn" T-shirt and was photographed next to my childhood hero, Mickey Mantle, whom I met once.

Gerry Holum and his wife Joanne own the remarkable Wagner Standard barn described in the Standard barn chapter. This structure that stands on their land near Hellertown is the Babe Ruth of Standard barns in the hearth area. This is one structure that has layers upon layers of extraordinary features and is fit for inclusion in the Barn Hall of Fame.

To two friends who both know how important barns are to me: Kathleen Keady, who accompanied me in 1993 to the Oliver Dutch-American barn in the mid Hudson River Valley in New York state; and Blyss Hoff (MD), who takes backcountry roads to see things in much the fashion as I do.

Also, I want to thank my two editors, Cheryl Weber and Alex Potter, who gave me many helpful suggestions along the way. Thanks to Roger Yepsen who first looked at the manuscript, and to Monique Reddington who formatted the text.

Finally, my hat is off to the anonymous North American barn builders of the 1625–1850 era, as well as the people who cruised old-growth timber forests with an eye for the best trees with which to create the great early timber-framed barns. You saw things I can only dream about.

# Introduction
# Places of Discovery

The Bilhuber house in Bergen County, New Jersey, is a two-century-old Dutch-American house with classic, local-style gambrel roof proportions. I first visited here in 1953.

In *American Barns and Covered Bridges*, Eric Sloane made a metaphysical connection between ancient Greek ruins and remnant barns. He wrote, "I feel that I am in the presence of greatness when I happen upon a crumbling ruin of an early barn" (Sloane 1954). Such a view could only come from a man who had deep connections to old farm buildings. He said many thought-provoking things about barns in his writings. One of the most poignant observations occurs in *An Age of Barns*, where he wrote, "It takes only an instant for a person to be directed to a path that he will follow for the rest of his life" (Sloane 1966).

My association with barns probably had its primal beginning in September 1953 in the small borough of Maywood, New Jersey, a few miles west of New York City.

At noon, after school let out, my Dad, John Huber Jr., picked me up where I waited for him on the front stoop of an early nineteenth-century Bergen County Dutch-American sandstone house. That house had a pervasive historic aura, and whenever I pass it, I still get the chills. More than 225 early Dutch stone homestead houses dotted the Bergen County landscape then, and their effect on me remains undiminished more than sixty years later.

The use of wood also figured prominently in my awareness of early material culture. My dad created unique pieces of redwood lawn furniture, and the tang and sight of that cut wood is still fresh in my mind. In my early college days, the two of us made a trip to the Whittingham sawmill in bucolic Sussex County, in northwest New Jersey, for several long, wide planks of walnut. The wood had a presence that conveyed a pervasive sense of both history and the rural countryside. Several years later I used those planks to make coffee tables, with gnarled branches from a great massive oak for legs.

Soon after, I had my first encounter with historic timber frame construction. Dad and I traveled to Orange County, New York. He had agreed to help a friend do some electrical work on his mid-nineteenth-century frame house. As he poked various holes in walls, I glimpsed at what he identified as hand-hewn beams, and I was intrigued.

Fall 1970 was the turning point in my connection to historic buildings and timbers in old structures. My family vacationed in Lake George, New York, in the Adirondacks, where a motor lodge included a number of cabins constructed of softwood logs. That, combined with seeing large-diameter native white pines, gave me a sense of what woodlands must have been like 250 years earlier.

Another early pivotal experience was seeing the Sierra Club publication, *The Last Redwoods and the Parkland of Redwood Creek*. I was struck by the fact that before the white man arrived in North America, much of the land was covered by a nearly continuous expanse of trees—and not just the West. I realized that trees likely had been 3 to 4 feet across, right where I lived in northern New Jersey. I later learned that these trees provided timbers for the region's first- and second-generation barns. Soon after that, I purchased *The Log Cabin in America: From Pioneer Days to the Present*, in which C. A. Weslager included a long chapter on the Pennsylvania Germans and Scotch-Irish. It included many illustrations of log houses. I began to appreciate how timbers from forest resources were utilized in the production of some of the earliest buildings in the eastern US. These houses were my introduction to the broad array of North American vernacular building traditions.

About 100,000 acres, both public and private, remain of old-growth redwood forests out of an original two million acres in California.

Tree identification guidebooks, such as *Know Your Woods*, indicated that the two most important species in the eastern US were oak and pine. Besides having a

Matson Woods, east of Bethlehem in Northampton County, is the best example of a very old second-growth oak community in the Pennsylvania hearth.

remarkable visual appeal, oak was a vastly superior wood product. This is why timber framers from the 1600s to the late 1800s often used oak for building construction in the northeast. Pine was used predominantly in northern locations such as the upper eastern river valleys in New York state and in New England.

I then fell under the influence of a remarkable book, Eric Sloane's *A Reverence for Wood.* His strong line drawings and text changed my life. The book chronicled how trees were used to produce myriad objects.

After obtaining a degree in biology, I began looking into forestry school. During two semesters at SUNY in Syracuse, New York, I was greatly inspired by my professor, Edwin Ketchledge, and came away with not just a greater regard for the importance of forest resources, but a heightened sense of their aesthetic qualities as well.

About that time I saw an ad in a local Bergen County, New Jersey, newspaper for a barn in upstate New York, that could be dismantled for $1,800. The owner gave me a tour of his mid-nineteenth-century barn, and I was delighted to see so many new things. The parallel saw marks on pine siding particularly fascinated me. In my imagination, I could see loggers felling immense trees and hauling them to a local sawmill powered by a mighty stream. I chose not to take on the big job of salvaging the barn (happily, that barn still stands), but I was moved by this tour of a barn on its original site. I knew then that the simple things standing before us have layers of history.

It wasn't long before I met Sloane at the Museum Village at Old Smith's Clove in Orange County, New York. I talked with Sloane several times after that and was pleased to be invited twice to his house in Warren, Connecticut. As we conversed, huge timbers from a relocated Canadian barn hung overhead, perhaps making sure that we focused on the topic at hand.

I began to visit barns regularly within roughly one hundred miles of my house. In Orange County, English-style side-wall-entry Ground barns were plentiful. They clearly conveyed the essence of pioneering so poignantly illustrated in the color plates of Sloane's *An Age of Barns.*

Another major vernacular barn style I observed was the Dutch-American three-aisle end-wall-entrance structures common in eastern upstate New York and the northern half of New Jersey. Several were almost literally in my own backyard.

At the same time, I was introduced to the two-level forebay bank barn found almost everywhere in southeast Pennsylvania. Massively built, they had a commanding presence; many were stone, with the largest ranging from 75 to 100 feet long.

Another major style I experienced then was the so-called swing beam barn with its single massive transverse tie up to 2 feet tall. These remarkable structures were found in several areas including the eastern river valleys in New York state and especially in west-central New Jersey. I later discovered that the style was also built in certain areas of the Pennsylvania hearth as early as the 1780s.

I was riveted by the different ways in which timber framers and barn builders in different areas responded

The circa 1760 Wortendyke Dutch-American barn in Bergen County, New Jersey, is one of the oldest remaining barns of its style in the East. The author gave his first barn talk here in 1977.

The DeLong double log crib Sweitzer in northern Berks County was built circa 1810 and is in excellent condition.

The *liegender Dachstuhl* roof style is the oldest known pure roof structure in all settled German areas in the hearth. More than fifty-five barns have been identified with this roof style.

to their particular geographical environments—something one is apt to miss by only concentrating on one style and place. By taking a broad view, it is possible to appreciate intimately what constitutes a true ethnic building expression. For example, by studying English barn style elements, the special qualities of Dutch or German barns become more readily apparent.

Around the 1976 bicentennial I made an important acquaintance with Robert Ensminger, who lived in Lenhartsville in Berks County, Pennsylvania. Bob, a professor of geography at nearby Kutztown University, was doing barn research basically by the seat of his pants—knocking on doors and obtaining permission to see barns and the wonders they possessed. Comparisons started to accumulate.

Over the course of more than fifteen years, Bob's fieldwork took him to several US states and Switzerland, where he established that eastern Switzerland was the birthplace of the distinctive forebay barn prevalent in Pennsylvania. This barn and its subtypes are featured extensively in his book, *The Pennsylvania Barn: Its Origins, Evolution and Distribution in North America.*

The first barn trip Bob and I made together, in May 1991,was the beginning of a close alliance that has taken us to several counties in southeastern Pennsylvania over a twenty-five-year period. Each of us saw things that the other might have missed.

I had made another important contact in 1988 when I visited the Old Stone Fort Museum in Schoharie County, New York, to view an elaborate display of timber bents from Dutch-American barns. This show was sponsored by the Dutch Barn Preservation Society, then based in Rensselaer, New York. I thought that here was a group dedicated to studying barns of a particular type and area. I soon joined that band of barn people and become part of their barn-saving efforts, in the process learning many things about timber framing.

That museum visit started me on a new track of looking critically at historic houses and barns.

After several years of intensive research on Dutch-American barns, I produced and edited the first two volumes of the *Dutch Barn Research Journal.* This may have been the first barn research publication ever produced and helped crystallize my thoughts. A few years later, I produced and edited the third and fourth volumes of the barn journal.

In the late 1990s, I updated the first edition of John Fitchen's 1968 *New World Dutch Barn* book and made extensive comments on his assertions. This second edition was titled, *The New World Dutch Barn: The Evolution, Forms and Structure of a Disappearing Icon.* This effort earned the Allen Noble Book Award, issued by the Pioneer America Society in 2003, for the best edited book on material culture in North America.

This prepared me for my move to Pennsylvania in 2000. I had developed the mental dexterity for sorting out the many expressions of barn-building manners.

When I made the move, Ensminger commented, "You have hit the mother lode." Was he ever correct! So many more barns occupied the landscape in southeast Pennsylvania than I had ever come upon in the river valleys in eastern New York and the northern half of New Jersey. Easily explained, as agriculture was king

The Duderstadt *fachwerk* Ground barn that resides in northern Somerset County, New Jersey, was built circa 1800. It has an appearance straight out of southwest Germany. Exterior *fachwerk* is covered by modern protective siding.

This Sweitzer near Schaefferstown in Lebanon County has a distinctive asymmetrical roofline. This early nineteenth-century structure has stone-to-the-peak construction at each end wall.

in the state. The fact that so many of the barns were constructed of native quarried stone made them seem, as folklorist Henry Glassie so aptly put it, like "the cathedrals of America."

I concentrated on the ten main counties in southeast Pennsylvania east of the Susquehanna River and south of the Blue Mountain, as these were the state's earliest settled areas. I later came to call this the hearth area of Pennsylvania.

I started to fill notebooks with architectural documentations, and in November 2001, I started a historic and cultural resources company called Past Perspectives. This exposed me to a wonderful array of homesteads in all the counties in the southeast part of the state and gave me a deep sense of the settlement patterns of the few ethnic groups that put down roots here, the barns they built, and the relative concentrations of barn types.

My work resulted in several articles. One thirty-page article, "Behind the Threshing Doors: An Inside Look at Some of the Earliest Barns in Pennsylvania," included a rare 1754 (dendro-dated) Lancashire-style Ground barn in Chester County that inspired me to compile a list of the earliest surviving barns and barn remnants in the state. This work helped me understand the evolution of barn-building traditions. Later, I made a list of Pennsylvania's dated barns per decade built prior to 1820. This study, too, revealed which dated barns had survived and the changes that occurred across several decades in timber framing.

Continued fieldwork resulted in two other important articles. "Barn Roof Timbers that Helped Build Early German-American Culture in Pennsylvania: *Liegender Dachstahl* and Other Roof Supports," involved examinations that brought me into direct and exclusive contact with the earliest methods of German roof timber framing technologies of the mid to late eighteenth century and early nineteenth century. It demonstrated both the consistency and variability of roof framing methods of certain German timber framers from 1750 to 1840.

Another article, "From the Distant Past: Early Type *Fachwerk* German Ground barn in Central New Jersey," concerns the important discovery of a rare barn in, of all places, northern Somerset County, New Jersey, a heavy Dutch-American settlement area. There, at the Duderstadt farm, probably in the 1790 to 1810 era, a German family and builder constructed a traditional Ground barn that looked like it came straight out of the Palatinate in southwest Germany. This barn has a remarkable number of classic early German building features, including the *fachwerk* or half timbering that was prevalent in medieval Europe. Its exterior walls are temporarily covered with protective wood siding. The barn is the single most outstanding example of its construction type in North America. More of its ilk may have appeared in the hearth area. Its presence may be further indication that the architectural landscape of the 1750 to 1850 era was far different than what is seen today.

In 2007, a farmer named Sheila Miller in Host, Berks County, started a state barn preservation group that I joined. We created a survey form to be completed by barn owners and interested others. The group, called the Historic Barn and Farm Foundation of Pennsylvania, grew from its original four members to 150 members. Since 2008, barn tours and lectures have been offered at annual June meetings, and the group has produced several barn tourguide booklets that tell the story of about fifty barns.

In 2010, I initiated the Sweitzer-Log Barn Survey of North America. It sought to locate and identify as many Sweitzers and log barns in North America as possible. Sweitzers occur almost exclusively in the eastern half of the United States. Log barns occur in many states of the Union plus certain areas in Canada. It is thought that 1,000 to 1,500 Sweitzers may survive on the continent, while log barns likely number in the several thousands. Both types are among the earliest barns remaining on the continent.

In the past forty years, traditional timber frame builders have also helped raise awareness of the significance of old barns. These craftsmen have learned the techniques commonly used in the eighteenth and nineteenth centuries. Prior to about 1970, such craftspeople were rare. Thanks to these dedicated workmen, many hundreds of old barns have been preserved.

## About the Barns

This book focuses on barns of Pennsylvania's hearth area, which saw the establishment of identifiable barn-building traditions in the eighteenth century and even earlier—in the era before 1650 when Swedes and Finns were infiltrating the region. In his 1989 groundbreaking book, *The American Backwoods Frontier: An Ethnic and Ecological Interpretation*, Terry Jordan states that the "Schuylkill route" (following the Schuylkill River) " . . . became one of the major arteries of American migration to the western frontier, connecting the Delaware River core to the Great Valley of the Appalachians." He added, "That Swedish and Finnish settlers led the way up this route and stationed themselves along it is highly significant to the evolution of backwoods pioneering." (Jordan 1989) He finished by saying that although the barn-building customs that prevailed more than 300 years ago will never be known, they profoundly shaped eighteenth-century barn-building traditions. Remember, too, that the conventions themselves changed over a one-hundred-year period. Each tradition was replaced with a newer tradition by successive generations.

After studying old barns for forty-plus years, I have come to realize that only a limited story of barn building between 1625 and 1875 will ever emerge because the builders are not alive to tell the story. Their ideas and decisions regarding style, design, construction, and fabric are lost forever. Our level of intimacy with the forces that created these barns is meager indeed.

However, this need not stop us from deeply appreciating their building methodologies, and we may intuit some sense of the spirit in which the builders did things. Perhaps that is why Sloane used the word "reverence" in the title of his 1965 book, *A Reverence for Wood*. Sloane knew that barns are more than the sum of their physical parts; he knew that the original builders imparted their own souls in the making of barns.

The Pennsylvania hearth lies south of the Blue Mountain and east of the Susquehanna River. This area, which covers many thousands of square miles, includes eleven counties: Bucks, Montgomery, Philadelphia, Delaware, Chester, Northampton, Lehigh, Berks, Lancaster, Lebanon, and Dauphin. This section entails the earliest settled geographic areas that influenced a great diversity of cultural manners that diffused to many areas beyond the hearth or core area. Ideas spread from a few groups to others. The Pennsylvania hearth must be considered one of the main hearths in the eastern part of North America. The barn-building traditions, in particular, established in the hearth during the eighteenth century, principally those of the German-American and English-American cultures, are responsible for many of the building conventions that spread far beyond this area. This included areas out to the Midwest and beyond and south to Maryland and Virginia, especially the Shenandoah Valley and into West Virginia and elsewhere in the south.

## Methodology

My approach to locating the barns in this book was randomly based. I traveled hundreds of roads in the eleven hearth counties but did not attempt to record each barn. Nevertheless, since literally thousands of barns have been documented, my observations are accurate and made with care. This is not to say that the way I've interpreted barn-building traditions are set in granite. Far from it! Each scholar makes his or her own conclusions.

More than 20,000 barns remain in the hearth area. About 5,000 barns have been studied to various degrees. One exception to the hit or miss approach is the road-by-road survey I took of barns in Lehigh County that yielded nearly 1,700 barns. Another exception is the Longswamp Township, Berks County, barn survey I conducted, which resulted in 106 barns. A final systematic survey involved roughly twenty barns within a quarter mile on either side of Route 100 from the juncture of Routes 100 and 29 in Hereford Township southwest to Bally in Berks County.

The more barns you see and the closer you observe their various aspects, the more you will understand that they contain layers and layers of information. It is an ongoing adventure—a world of discoveries. Perhaps this book will inspire you.

Finally, I hope this book will dispel the belief that "barns are just barns." Like anything else, these structures have a long history of evolution. Their appearance on cultural landscapes depended on an untold number of men and materials. Barns have a web and weave all their own. This is a story that, while certainly incomplete, will hopefully fill some of the gaps. As Sloane said nearly fifty years ago, barns "are rather the shrines of the good life and ought to be remembered" (Sloane 1966).

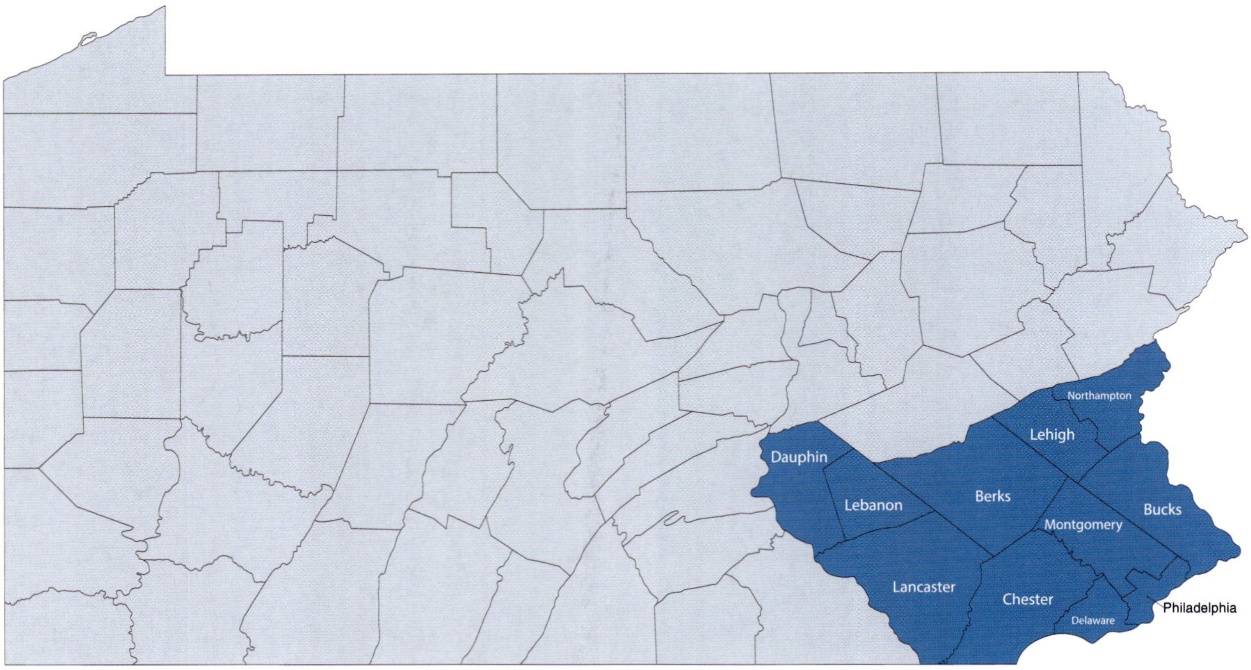


**Pennsylvania Hearth Area Map**
(Eleven counties in blue)

"Nature has no greater triumphs than its forests, trees, and wood."—Anonymous

# Chapter One. Europe before the First American Barns

By studying the history of architectural expressions in Europe, we can better understand how North American building traditions first became established in the seventeenth century and evolved in the eighteenth and nineteenth centuries. European settlers in North America brought with them many vernacular building traditions. Just as a dialect is distinguished by its vocabulary, grammar, and pronunciation, vernacular buildings have their own style, design, construction methods, and fabric. Such buildings, especially those within close-knit communities and well-defined time frames, can vary strikingly from one region to another, even across relatively short distances. Countless buildings, one distinguishable from another, appeared on landscapes across Europe from medieval times to the seventeenth century and beyond.

Enough barns remain that we can get some idea of how building methods and materials evolved. This lexicon of architectural features amounts to a pattern recognition system. We know, but only to a degree, what builders knew hundreds of years ago.

In his classic book *The Log Cabin Myth*, Harold Shurtleff explains how this "transit of civilization" works: "... as a large body of people that possesses a relatively high culture ... moves to another and distant area inhabited by people of a different and (to the newcomer) lower civilization, the emigrants attempt to preserve in their new world as much as possible of the manners, customs, folkways, language, literature, arts, and crafts of their former home" (Shurtleff 1939).

Europeans brought three main barn types to America: the English and Dutch styles, which originated in the seventeenth century, and the Swiss-German style from the eighteenth century. *The New World Dutch Barn*, by John Fitchen demonstrated that the basic three-aisle ground form of the Dutch-American barn was borrowed directly from certain areas of the Netherlands and possibly from lowland West Germany. Mid- and late-seventeenth-century surviving Dutch contracts testify to this assertion. And in *The Pennsylvania Barn*, Robert Ensminger showed that the Sweitzer form of the forebay bank barn came from eastern Switzerland. One major North American barn type whose origin has never been authoritatively documented is the English side wall wagon-entry one-level Ground barn. However, given elements such as upper tie beams and gun-stocked (flared) side wall posts, there is little doubt that this barn originated from English prototypes.

## Building in a Deforested Europe, 1200 to 1600

The earliest North American barns were made of forest timbers, and a knowledge of forests will help us understand why barns came to look the way they did. Around 1600, North America was largely a wilderness with many areas covered with trees 2 to 4 feet in diameter (Braun 1950). By comparison, England at that time was only ten to fifteen percent forest, as most of the land was inhabited, according to the *Domesday Book* which documented the great 1086 land survey of the English countryside. The massive timbers (up to 20 inches across) that went into the construction of the great cruck-built tithe barns, for example, such as Leigh Court in Worcestershire, England, were almost unobtainable after 1500.

In *Conservation of Timber Buildings*, Fred Charles notes that large timbers in Europe were in short supply during the medieval ages. For centuries English barns were built with small and often crooked sections of wood. Timber framers in Europe had to meticulously "pick and choose" the trees they worked with. When they needed large timbers, they often had to travel considerable distances.

The rise of the iron, pottery, and glassmaking industries are blamed for the destruction of many forests. However, farmers with their ever-present plows caused more disruption to sylvan environments than any other single force. Wanting to make long, straight rows for their plantings, they would not just cut down trees, but in many cases would grub out the stumps. About half the woodland in lowland England was destroyed in the 700 years between the initiation of the Iron Age and the Roman occupation. This condition also likely existed in other European countries such as Germany and Switzerland. These are the "forests" that immigrants to America in the 1600 and 1700s knew.

Until the last century, people typically did not have a sentimental connection to trees because they depended on them in ways that today we do not. Yet they venerated trees. They knew how often oaks produced leaves per year. They knew that young oaks would sprout again and again even though animals would eat entire shoots. They used every part of oak trees, from wood to the leaves to the bark to the acorns to the galls. In *Oak: The Frame of Civilization*, William Bryant Logan states, "the Druids served as lawgivers and singers. The honorific name of *Druid* is derived from two words: dru, meaning 'oak,' and 'wid,' to see or to know." The people who were thought fit to sing the tribe's stories and to interpret its laws were said to "know the oak." These ancient people honored oaks perhaps as no other humans ever did. It was especially they who knew the greatness of these trees as they directly experienced them (Logan 2005).

A poignant story of the difficulty medieval people experienced in sourcing properly sized timbers involved Abbot Suger, a regent of King Louis VI who lived circa 1081 to 1151. Suger was one of the last Frankish statesmen, and was also a historian and the influential first patron of Gothic architecture.

In erecting the Saint-Denis basilica, a large abbey in the commune of Saint-Denis, a northern suburb of Paris, Suger needed about a dozen substantial trees for the roof structure. Roland Bechmann, in his superb book *Trees and Man: The Forest in the Middle Ages*, relates that Suger consulted local wood workers in Paris, who told him that no trees of the dimensions he needed were in any local forests. He left the area with his carpenters and the dimensions of trees that he needed in his mind. He went on and on until he found the appropriate trees for the basilica. Apparently he would not settle for anything but the best trees for the job at hand. He was imagining the steps necessary, however difficult, in creating an awe-inspiring work of architecture.

As Otto von Simson said in *The Gothic Cathedral: Great works of Art*, "Great works of the human mind are always called forth by what Mr. Toynbee [the historian]has defined as the law of challenge and response" (von Simson 1956). The builder accepted the dare of the building requirements and answered by going the extra mile to find large timbers. The time spent was both a function of the great scarcity of large trees and the dedication of the builder.

Another story concerns the search for great trees for the 65-foot-span, double-hammer beam roof of Westminster Hall in London that began construction in 1395. Appropriate timbers were found near Farnham about forty miles southwest of London. Twenty-six half-beams and twenty-six pendant posts were carried in two carts with sixteen horses that made fifty-two journeys to Hamme on the Thames—one journey for every piece of timber (Charles 1984).

## Coppicing

In Europe, it is likely that the limited timber sizes influenced elements such as roof structures and bents. In his treatise "A Brief History of Coppicing," Edward Mills states that the shortage of large trees led to the practice of coppicing, a woodlot management system going back as far as the Stone Age. Early Greeks and Romans used it extensively, as it was the only form of forestry they knew. In France and Germany, coppice was practiced during the Middle Ages mainly for the production of fuel. The French called the practice *bois revenants* or *bois de vente*. In Germany it was known as early as 1359, and in Switzerland coppice regeneration was first used around the sixteenth century.

Coppice comes from the French word *couper*, to cut. When trees were felled, hardwood trees regenerated themselves making so-called coppice shoots. Rotation periods of five to twenty-five years were practiced and cuttings occurred almost invariably in the winter at ground level. Underwood, often used as fuel, was the most important crop of English coppicing. Part of this management system yielded standards, or timber-sized trees, more than 8 inches in diameter and 18 to 20 feet tall before the first branches grew out of the tree trunks.

It is perhaps no coincidence that the branchless lengths of standards determined the width of many domestic English timber-framed buildings. Indeed, the work of Oliver Rackham in assessing the use of medieval oaks in buildings of that era affirms this. He contended that all but the grander buildings are about 20 feet wide. Further, he believed that "ordinary medieval buildings contain large numbers of small oaks felled usually at 25 to 70 years of age." For example, a close examination of Coppard's Bridge, an open hall house in the Low Weald seven miles north of Lewes in England, revealed that 111 trees were used in its construction. Only four beams (oak) found by dendrochronology had more than fifty annual growth rings while the most had just seventy-eight rings (Peter Varlow 2016).

Tree lengths greater than 20 feet were certainly obtainable in the English countryside but these were reserved for public buildings such as cathedrals and other grand buildings. In coppices of England, standards were common. Dimensional timbers were obtained for house and shipbuilding. *Quercus robur* or English oak had a rotational cutting time of 120 to 140 years for construction timber. Hedgerow standards also provided building materials.

Another means of producing shoots involved a process called pollarding. When tree shoots were cut 6 to 8 feet above ground, the trees were called pollards. Shoots in certain cases were allowed to grow to widths that were suitable as timbers for particular buildings.

It bears repeating that these timber management practices were in distinct contrast to what the immigrants experienced after their long trek across the Atlantic Ocean.

## Tree Resources in England—Medieval Times Versus Today

The sizes of trees available in modern times versus what was generally available in medieval times in Great Britain is significant. A situation in 1970 in York well illustrates this widespread condition. Fred Charles, a preservation architect, had to locate six beams for the restoration of York Minister tower near the River Ouse. The decayed original timbers were just over 50 feet long and about 2 feet square. Just one tree was found despite the fact that every timber merchant in the country was contacted. The circumstances necessitated that steel lattice beams had to be used.

The dimensions of trees available to builders 600 to 900 years ago were of paramount importance and helped to determine many aspects of structures built then in the Middle Ages. Lyn Courtenay, in her treatise of the evolution of the aisled hall to the un-aisled hall, says this about one important building type of that time—cruck structures. "Full cruck spans are naturally limited by the size of the trees available and rarely exceeded 10 meters; and yet impressive halls such as the four-bay hall at Stokesay Castle Shropshire [in England], dated by dendrochronology to 1285 and spanning 31 feet internally, survives to illustrate how these massive timbers might be employed where the framing begins some distance up from ground level [and in fact, the original blades have been truncated]. Generally, however, span of half that distance or less would be more of a norm for ordinary houses." These "massive cruck timbers" were by no means readily available back in the late thirteenth century. A cruck (blade or crook frame) is a long, curved timber (one in a pair) that supports the roof of a building. Such timbers lean inward and form the ridge of the roof.

The British dendrochronologist Doctor Daniel Miles says in his 2006 article, "Refinements in the Interpretation of Tree-Ring Dates for Oak Building Timbers in England and Wales," that "the average age of the oak in a medieval building is between 25 and 100 years." He continues, "The general rule was to use as small a tree as possible to get out the structural member required." Here we have virtual proof that large trees (over 2 feet in diameter) were generally scarce.

This is a basic look at what was then obtainable for many builders in England, in contrast with the forestry conditions the British experienced in America starting in the first half of the seventeenth century.

## Tree Resources in Switzerland and Germany

Many of the immigrants who built barns in Pennsylvania's hearth area came from Germany and Switzerland. In Switzerland, land use practices also indicate an awareness of the importance of forest resources. Schmidthüsen and Zimmermann (1999) state:

> "Of particular importance was the medieval colonization with Land Clearing in the Central Plateau and the Jura. This peaked during the twelfth and thirteenth centuries. Forests in the vicinity of villages were intensively used for local wood supply, food, and other forest products, and as part of the agricultural space in combined production systems. From the 15th century onward large-scale forest exploitation for mining, salt production as well as for artisanal and industrial uses developed."
>
> They continued:
>
> "An interesting aspect in the Swiss forest history is the collective and communal land use systems and the logging bans for forests with important protective functions. They have been common in the Alps and Pre-Alps since the 14th century. In the Alps common resource use systems have been maintained and developed into communal and corporative forest ownership. In other parts of the country, forest areas were divided among their owners and became private farm forests. Already during the 17th and more strongly in the 18th century the reduction of usable forests and the fear of a general shortage of wood induced efforts towards a more sustainable utilization and management."

These observations point to the fact that anything remotely close to the abundant old-growth stands of timbers in eastern North America were long gone by the time Europeans immigrated to the US.

Germany's forests also changed dramatically over a long period of time. The Hercynian Forest was an ancient and dense forest from the Rhine River across southern Germany. The Black Forest, which extended east from the Rhine valley, formed the western side of the Hercynian. Across the Rhine to the west extended the *S*ilva Carbonaria mountain ranges and the forest of the Ardennes. These old-growth forests of ancient times were the original post-glacial temperate broadleaf forest ecosystem of Europe.

In his well-known treatise, "Natural History, "Pliny the Elder wrote that the eastern regions of the *Hercynium jugum*, the "Hercynian mountain chain," were in Pannonia (present-day Hungary) and Dacia (present-day Romania). He provided an elaborate discussion of its composition, noting that the closeness of the trees resulted in their severe struggle to survive. He does not fail to mention the great oaks, and his prose reflects his thoughts and feelings about "the gloomy forest."

Many connected forest areas were destroyed as trees and were used extensively for firewood. In the Black Forest, huge quantities of timber were tied together to form rafts and exported to the Netherlands, where the wood was needed for shipbuilding. This activity acutely depleted forests. Farmers over-exploited forests for centuries by means of grazing and in the making of straw. At the end of the eighteenth century, there were hardly any forests left in Germany. Wood eventually became so scarce that, in winter, fence posts, steps, and all kinds of wooden objects that were expendable in the short term were burned as firewood.

It is apparent that in areas of both Switzerland and Germany, magnificent forests existed centuries ago. By the time of the first half of the eighteenth century when extensive migrations to America occurred, the great forest resources were much diminished. When Swiss and Germans went to America, the forests they experienced were almost nothing like they knew in their homelands.

# Chapter Two.
# Early North American Barns

The Ono double log crib Ground barn in the Ono area of Lebanon County was built in the late eighteenth century. It was dismantled in the late 1970s and awaits reconstruction. (Courtesy of Robert Ensminger)

Barns of one kind or another have been built in North America for a little more than 400 years. Many scholars consider Jamestown to be the first successful permanent settlement on mainland North America. The area was founded in the Colony of Virginia in May 1607. Only thirteen years later, in 1620, Plymouth Bay Colony (present site of Plymouth, Massachusetts) was settled. No one knows when the first-generation barns appeared. They may have housed farm crops or stabled farm animals or both under one roof in Massachusetts and Virginia. None of the first-generation barns in Virginia or any other English settled area apparently survive.

It seems that the earliest constructed barns were fashioned in New England, New York, and Virginia. Archival sources indicate barn dates as early as the 1630s in these areas.

Regardless, the European settlers must have been delighted to find a nearly infinite supply of trees that could provide 2-to-3-foot-wide timbers for barns. They found abundant large hardwoods such as oak and ash and softwoods such as pine and hemlock. Builders in the Northeast particularly favored oak; pine was more popular in the North.

## Barn Forms and Barn Styles

It is important to distinguish between barn forms and barn styles. Barn forms are distinguished as one-level, two-level, and three-level structures. One-level barns are normally referred to as Ground barns or flat barns. In probably all areas of North America, first-generation barns, and perhaps many second-generation barns, took this form. These barns did not include basements, at least not below their wagon bays.

Two-level barns are usually banked structures with dug-out sections for basements; loaded hay wagons could enter through wagon doors.

Another type of two-level barn had a built-up earth ramp adjacent to the rear-wall wagon doors where the wagons could "climb up" the earth slope and then enter the barn.

Rare exceptions exist of two-level barns that are unbanked, such as at the early Keim homestead in Pike Township and at the Kaufman homestead in Oley Township, both in Berks County.

The third barn form is the Double Decker, or three-story structure, which appears in several states. Hay wagons traversed an earth bank and then crossed a bridge to enter the barn. Animals were stabled on the ground floor, while the top floor was for threshing and had mows at each side. A middle level housed

The DeHart stone Standard barn in Berks County near Reading is an 1861 structure. Unlike many Standard barns, no house-side basement door is seen on the end wall.

This small three-bay stone-to-eaves-level Standard barn in Upper Milford, Lehigh County, is painted an unmistakable white. The barn has a built-up earth ramp to the wagon doors.

The Keim homestead in the Oley Valley, Berks County, possesses a rare two-level barn without a bank. It has a forebay on each of its eave walls, making this a non-Pennsylvania barn.

A stone Double Decker of rare serpentine stone north of West Chester in Chester County was built circa 1820.

sink mows at the sides, and grain bins in the middle, for threshed crops. Few Double Deckers were built prior to 1800.

A barn style, on the other hand, relates to its construction method, materials, or regional character. Most barn architecture evolved to reflect changing design ideas and technology.

## Four Major Early Barn Styles

Four major early or pre-1775 North American barn styles have been recognized in barn literature in the past sixty years. They include English side-wall-entrance one-level barns, Dutch-American three-aisle barns, swing beam barns, and Pennsylvania forebay bank barns. Three of the barns, (excluding the swing beam barn) were discussed and illustrated in Eric Sloane's *An Age of Barns*. He gave us a proper sense of what these main barns looked like, but provided no information on when the first barns of each culture area were built. Another barn style—the German Ground barn or *grundscheier*—might be added to the list with a few others. Within each of these barn styles are various barn types; some of them are discussed in separate chapters.

Descriptions of these barns are included here to provide a comparative look at barns built in the Pennsylvania hearth and in other states.

**English Side-Wall-Entrance One-level Barns.** A few sources in the New England Historical and Genealogical Register refer to barns built in Massachusetts as early as the 1630s and 1640s. A few specimens of late-seventeenth-century English barns, or at least remnants, have apparently survived in Massachusetts. They are likely the oldest existing barns in North America.

The first English barns were probably one-level barns that consisted (usually) of three bays with a side-wall wagon entrance and measured approximately 30 feet by 40 feet. Side wall bents with distinctive gun-stock

This circa 1800, English-style side wall wagon entry one-level frame barn is in southern Vermont.

This New England-style frame barn has the distinction of having an off-center gable wall wagon entrance. Thousands of such barns, many of them joined to the homestead houses, were built between 1825 and 1925.

(flared at the top) posts and the attendant English "tying joint" were also present. The joint with post is a section of the frame where wall post, wall plate, roof rafter, and upper tie beam intersect. Many early English barns have such joints. The main English settlements were in New England and the areas surrounding Philadelphia, and sections south in the Delmarva Peninsula, an area of about sixty by 180 miles. Many thousands of these basic early English barns remain in North America and date from 1775 to 1850.

One of the earliest existing English barns is the Bardeen-Culver barn (a remnant) in Newbury, Massachusetts. This barn was much altered from its original state but the earliest timbers (dendro-dated) originate from the mid-1690s. The barn has the distinctive "tying joint" with gun-stocked wall posts. Another early specimen of an English barn that may date from the 1680s is near the Rhode Island-Massachusetts border. And a barn at the White-Ellery place in Gloucester, Massachusetts, has been dendro-dated to 1739.

To date, no authoritative book on English barns in an English settled area in North America has been published. The closest offering is the 1984 book by Thomas C. Hubka, *Big House, Little House, Back House, Barn: The Connected Farm Buildings of New England.* Another book, which appeared in 2000, is the *Field Guide to New England Barns and Farm Buildings* by Thomas Visser. This covers the gamut of barns across many eras and mentions a few late eighteenth-century barns.

A curious note appears in Harold Shurtleff's classic 1939 book *The Log Cabin Myth: A Study of the Early Dwellings of the English Colonists in North America.* Shurtleff quotes William Penn: "that if a person starts building a house of this sort [frame construction] about October 8, he will finish it by November 15, and the barn by the spring." This quote apparently refers to an English-style barn built in the hearth area of Pennsylvania. Did Penn mean that the barn was partially constructed in the fall and then completed in the spring? Or was the entire barn built in the spring? In any event, this statement that originated more than 300 years ago is a unique reference to the length of time required to erect a barn.

The Couvenhoven Dutch-American barn in Monmouth County, New Jersey, was probably built in the second half of the eighteenth century. A few miles from the Atlantic Ocean, the barn has early-type classic proportions.

Dutch-American Three-Aisle Barns. The earliest Dutch settlements were in the eastern river valleys in New York state in about twenty counties and the northern third of New Jersey in about ten counties. Dutch-American scholar Charles Gehring says the earliest reference of a Dutch-American barn dates to the 1630s and cites the house-barn of New York director Kieft, head of the New Netherlands in New York. Early or pre-1820 Dutch-American barns are also one-level structures with end or gable wall wagon entrances. They had a three-aisle format—a central wagon passageway and two side aisles. The barns possess the most distinctive roof silhouette of any early barn style in North America. They also have the diagnostic interior H-frames with often massive horizontal anchor beams. Such beams commonly were more than 18 inches deep in the Schoharie and Mohawk River Valleys in New York, as well as other areas including Monmouth and Somerset Counties in New Jersey. A few derivative barns of post-1820 vintage can be found in the Harrodsburg area of Kentucky. Of the three major barn styles, the Dutch-American barn has the most limited geographic area.

Less than a thousand Dutch-American barns survive. Several types of this barn were built, including several varieties in New Jersey. (I have identified 610.)

In the second edition of John Fitchen's classic work, *The New World Dutch Barn: The Evolution, Forms and Structure of a Disappearing Icon*, I supplied eleven references to seventeenth-century barns. This information originated from the fortunate survival of original contracts of that era. Five of the structures are the "pure barn" type—that is, the building stored only farm crops and provided stabling areas for animals. The earliest cited barn with dimensions was built in 1638. Two barns were from the 1640s, and two were from the 1670s. Other seventeenth-century barns may be found, including a few in Ulster County.

In addition, I refer to six so-called house-barns, combination structures that housed humans, crops, and animals. These *loshoes* were considerably larger than the conventional barns. The earliest cited house-barn was from 1634. Five structures dated from the 1640s. Two of the barns were massive at 100 feet long and 120 feet long.

Swing Beam Barns. This style is relatively unknown and has received scant attention. My article, "The Swing Beam Barn in the New World Dutch Cultural Hearth," was published by the Dutch Barn Preservation Society in spring 2011, and includes examples in several states.

A swing beam barn may be a one-level or two-level structure and has a characteristic large single transverse beam that borders the wagon floor. The beams can measure 12–24 inches in height. However, one example north of Toronto in Ontario, Canada, has a remarkable 35-inch height. The beams are placed about 7 feet above the floor. The critical qualifier of a swing beam is the absence of any posts or braces below the beam. No other barn style or type can make this claim.

Observers do not agree on the exact function of swing beams. One theory posits that horses (and other farm stock) were tethered to two-part wood contrivances (a vertical pole and a conical wood peg studded block) that were joined to the swing beam. Farm grains were spread out on the wagon floor and horses trod circular

This swing beam oak barn in the Cranberry area of Middlesex County, New Jersey, was likely built in the last quarter of the eighteenth century.

This Cranberry, New Jersey, barn has a swing beam nearly 20 inches high at mid-point. No posts, studs, or braces are seen below the beam.

paths or "swung" around and below the swing beam to separate the seed from the chaff. Another theory states that hay wagons were unloaded in the barn; the horses were then unhitched and the horses and wagon were swung around and below the swing beams to exit the barns. This may be why no posts or braces were used. It is possible that both practices were followed.

The existence of any pre-Revolutionary War swing beam barn has not been established. No one knows when this barn type was first constructed in North America, but it appears to be an American invention. Several barns from 1780 to 1800 occur in west-central New Jersey and southeastern Pennsylvania. A few barns built prior to 1800 may have been built in the upper river valleys in eastern New York. The Industrial Revolution, which dramatically changed agricultural economies starting in 1810 to 1820, seems to have been a distinct stimulus for the swing beam barn's proliferation after 1820. The barn was better adapted to farming ways in this more advanced cultural milieu. It was constructed well into the late nineteenth century and appears sporadically in eastern New York, west-central New Jersey, Pennsylvania, Ohio, Illinois, and Ontario.

Pennsylvania Barns. We know something about the earliest barns in the area that later became known as Pennsylvania, thanks to the pioneering work of Terry Jordan and his 1989 book, *The American Backwoods Frontier: An Ethnic and Ecological Interpretation*. He delved into the architectural contribution that the early Swedes and Finns made in the Schuylkill Valley—a major passageway of American migration to the west, and what Jordan calls the "Midlands backwoods culture area."

Jordan writes that the Finnish Savo-Karelians —"free-spirited, boisterous, alcohol-guzzling, party-loving backwoods folk"—made their cultural mark on this major communications route starting around 1640 (Jordan 1989). Their first-generation structures were made of round logs imperfectly brought together, and their manners were exquisitely pre-adapted to the extensive woodland environments they experienced in America—more so than just about any other group arriving in the seventeenth century. They were a significant segment of the population that constituted New Sweden after 1638. They first settled on the lower Delaware River in parts of present-day Pennsylvania, New Jersey, and Delaware.

That the Savo-Karelians carried some sort of notched corner log barn building tradition to America is basically unquestioned. They found that trees for construction were plentiful, but different from their familiar homeland environment of coniferous and birch forests. Instead, on the west bank of the Delaware were trees such as oak, walnut, chestnut, maple, hickory, and tulip (yellow poplar). The big trees gave way to their axes. At first, early observers relate that the settlers, then of New Sweden, built "*pirtii*-style cabins" out of round logs, consisting of one small room with a corner fireplace. But the records are mute on the subject of barns.

It is likely that first- and second-generation seventeenth-century builders used small trees or tree branches for construction. Small pieces of wood would have adequately supplied their needs, and their early tools would have limited their ability to work with large timbers.

Jordan believed that the "backwoods log architecture encompassed a small number of simple, efficient, and interchangeable house and barn plans. In houses, a four-log wall unit was a rectangular pen while outbuildings [including barns] had such units that were referred to as cribs" (Jordan 1989). He speculated that, in the greater Philadelphia area, house plans such as the dogtrot type (two cribs with a breezeway between) were accompanied by "apparently the open-runway double-crib barn as well" (Jordan 1989). This is the closest we come to knowing what Pennsylvania's earliest barns looked like.

The British also made their mark on architecture of the midland American frontier. They introduced the gable end wall fireplace in their houses, and it seems likely that they also influenced barn architecture to some degree in the seventeenth century.

What we can say is that the early New Sweden barns were one-level structures, and all available evidence, though meager, points to one-level barns as the earliest successors to the "Swedish colony" barns.

## Early-Eighteenth-Century Barn Documentation

We know that combination house-barn structures existed quite early as three of these building styles are referred to in Margaret Schiffer's 1984 book, *Survey of Chester County, Pennsylvania, Architecture: 17th, 18th, and 19th Centuries*. One combination building is mentioned in an August 1705 document generated at a Court of Quarter Sessions in Chester: " . . . a certain stable being part of the dwelling and mansion house . . ." In another case, the 1798 Direct Tax for East Caln and West Whiteland Townships lists Joseph Downing with "1 dwelling house and Barn Connec(te)d." The 1799 tax assessment for Honey Brook Township lists Abraham Curtz with "A Dwelling house and barn of Logs Covered with Straw and Shingles under one Roof." A 1705 combination house-barn was documented at the Unruh homestead in Germantown. In yet another example, the French naturalist and artist Charles Alexandre Lesueur (1778–1846) in the 1820s sketched a combination building of unknown date. This may be the only genuine depiction of a Pennsylvania house-barn structure in existence. Schiffer mentions several more structures of this early type, all of which must have been one-level structures. Such a building, in Germanic culture, is known as an *einhaus* (one house). It is quite likely as the earliest structures were of rather small dimensions, since banked or two-level structures would have been too ambitious in the first half of the eighteenth century. The early *einhaus* form resembles what came to be known as Ground barns in the hearth area.

How late these combination structures were built is unknown. Whatever evolutionary effect they had, they certainly added to the richness of early cultural landscapes. If a few barn styles existed in Pennsylvania before about 1725, what impact did these early barns have on barns such as the forebay barn style, first built in the mid-eighteenth century? It seems there was a great deal of architectural jockeying for position in the eighteenth century, and that Jordan's notion of "social winnowing" is on the mark.

Although there are few citations of barn structures in the first half of the eighteenth century, one outstanding exception appears, albeit briefly, in the article "Life in Bucks County in 1722/23," edited by Harold B. Hancock. It is one of the earliest references to a barn in Pennsylvania (not including an inventory or an indenture). A man

A Charles Leseur sketch depicts a rare combination house-barn structure near the Blue Mountain that clearly shows the house at one side and the barn with wagon doors at the opposite side. The building may have been built circa 1800.

named Ambrose Barcroft, living in Solebury Township in Bucks County, wrote a letter to his father in England about the trials and tribulations he was facing in America. Barcroft said he was " . . . building of a barn of 4 Bay. It would cost me near 30 pounds to let out, but the man I Just mention'd is something of a Carpenter, and I think with the assistance of a good workman a few days at the setting up the frame, He'll do the work. The Clapboard and Shingle I have bespoke and is already pay'd with goods that I have sold."

With the mention of the words frame and clapboard, we have incontrovertible proof that frame barns were being built in the first quarter of the 1700s. It is also interesting that the barn had four bays, which seems large for such an early barn. The house, curiously, was a log structure, as Ambrose testifies: "The walls of oke (oak) trees of about 12 inchis Diamiter, (was) laid one upon the other and let into another at the four corners." This is remarkable authentication that houses and barns in early times were not always built of the same materials at a given homestead.

The 2004 article "Behind the Threshing Doors: An Inside Look at Some of the Earliest Barns in Pennsylvania" was a review of a several surviving barns of the eighteenth century in Pennsylvania. It clearly stated that since few barns predate the Revolutionary War era, it is essentially impossible to obtain any precise description of barns built in Pennsylvania before about 1770.

The following are brief descriptions of the major barn styles that have been recorded on Pennsylvania landscapes since the publication of Alfred Shoemaker's book, *The Pennsylvania Barn*—the first of its kind, published in 1953. Each of the major barn styles and certain of their sub-types are discussed at length in separate chapters.

In this book, the attempt has been made to include all the frequently identifiable barn forms, styles, and sub-types that include the forebay barn style as a major component of barn populations in the hearth. Certain variations are not examined as they are much too numerous to include. A number of these "nondescript" barns are especially seen in the southern half of Montgomery County and to some degree in the southern half of adjacent Bucks County. This situation also exists to a fair degree in Chester County. Most of these unusual barns are two-level bank structures that do not fall neatly into the well-defined categories of the barn styles that follow.

## Ground Barns

The one-level or Ground barn was the first barn form likely built in the Pennsylvania region. It was presumably the dominant form in the seventeenth and early eighteenth centuries. Whether the barns contained both farm crops and stabled farm animals is not known. The basic form was built well into the second half of the nineteenth century, and hundreds built between 1780 and 1875 have survived on the landscape.

Both German and English-style Ground barns were erected. German-built barns were referred to as *grundscheiere* or *boddamscheiere* in the local dialect. Early ones were log crib construction. Some English Ground structures are referred to as Lancashire barns, named after an area in northwest England where they are found. These barns are scarce.

The Longswamp, Berks County, structure is a two-section Ground barn. The original section, circa 1840, at the left has two bays with one stone end bay; the later-built circa 1880 frame section at the right has one bay. The far left section is a late addition.

The Bertolet-Coker Sweitzer is a masterpiece of late eighteenth-century (1787) barn architecture and is replete with both rare and original features. The adjacent Standard barn was built in 1837.

This 1819 Miller barn in Maiden Creek Township is a finely crafted Standard barn. At the barn front (out of view) are the unusual and superbly built stone arches over the stable wall doors.

## Pennsylvania Forebay Bank Barns

This barn has been the focus of study since the 1950s. It is defined as a two-level banked barn with a distinctive cantilevered upper floor on the front. Three types of forebay barns have been identified.

## Sweitzers

Sweitzers with log, stone, and frame construction are from the eighteenth century—some are dated while others are believed to be of this era. Sweitzers were built, although in declining numbers, into the first half of the nineteenth century. Sweitzers have asymmetrical rooflines because of the distinctive forebays or frame appendages at the front. Sweitzers thus far have been identified in twenty-four counties in the state.

## Standard Barns

The Standard barn is the second recognized Pennsylvania forebay barn, first built 1790 to 1800 or several decades after the first Sweitzers came along. Only one eighteenth-century Standard barn has been identified—the DuPortail barn dated 1792 near Devon in Chester County. Standard barns started to proliferate after about 1820. This coincided with the beginning of the Industrial Revolution. After 1830, Standard barn construction was virtually everywhere, and a few thousand remain in the hearth area, dating from 1830 to 1900. Unlike Sweitzers, Standard barns have symmetrical rooflines.

The 1806 Standard barn is one of the earliest structures of its type in Chester County. As expected, it is stone to the peak.

## Extended Barns

The last and perhaps least-known forebay structure is the Extended Forebay barn. These were built between 1790 and 1920. Six sub-types of these barns have been identified. Extended barns provided more room to store crops. The most often encountered Extended Forebay barn type is the Up-country Posted Extended barn; a few thousand are seen in ten counties in the southeast part of the state. In almost every case, the front extensions or straw sheds were added to the main rear sections, most often Standard barns, after about 1875. Only a few of these barns are original in both the main sections and the front sheds.

## English Lake District Barns

This barn style apparently came directly from the Lake District in Northwest England. This barn most often has full front-wall stone construction, from ground to the top of the second floor level. The barn also has a full pent or front roof overhang at the front basement-level wall. This barn type is a true eighteenth-century structural concept that proliferated on English-based landscapes in the greater Philadelphia area. The barn style was likely built into the 1850s.

Examples of this classic Rockland Township, Berks County, Up-country Posted forebay barn populate most areas of eastern Berks County and the greater Lehigh and Saucon Valleys.

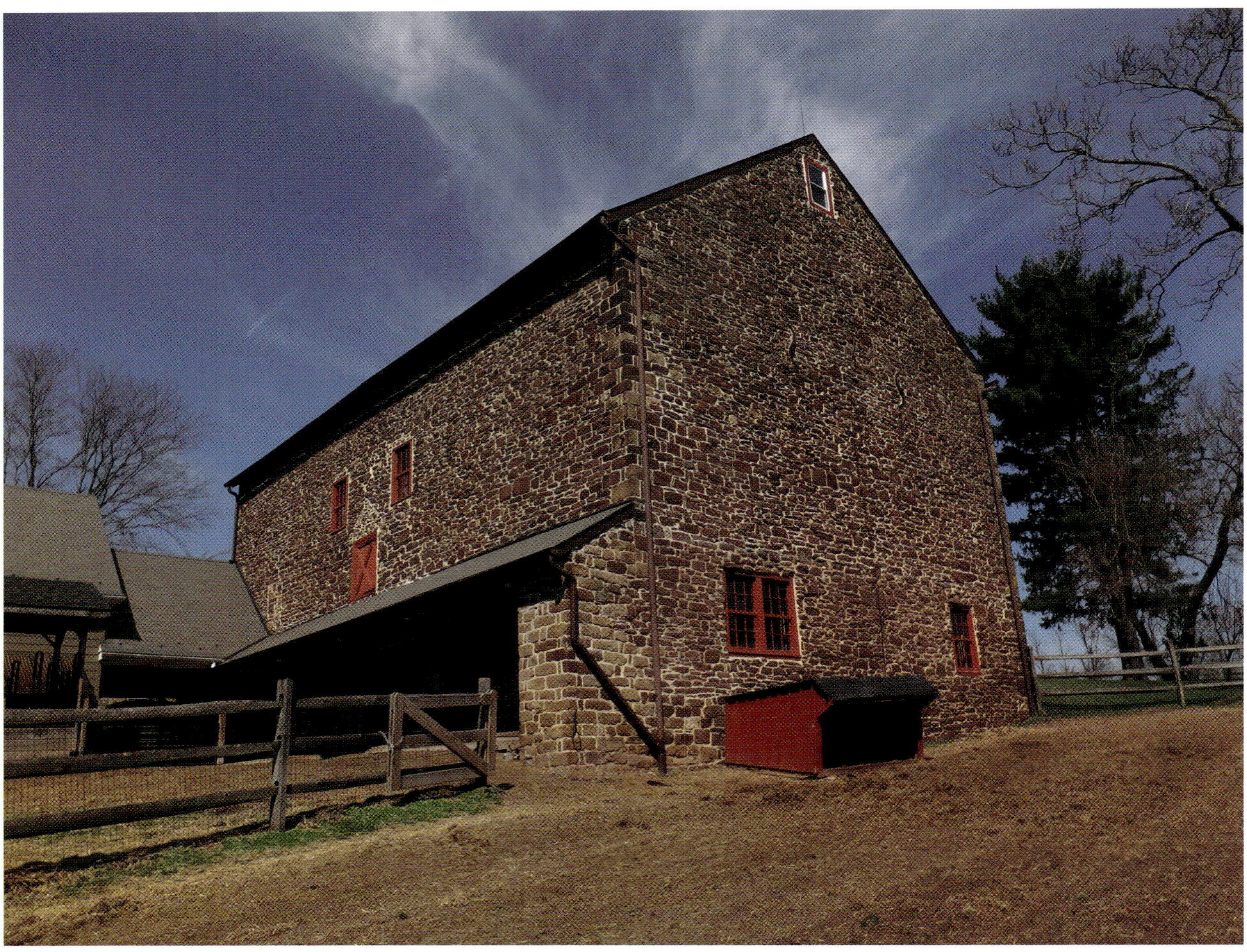

This circa 1825 stone-to-peak English Lake District barn is in northern Bucks County. Germans mostly settled this area and few of these barns survive in this general location today.

## Summary

Experimentation with both barn styles and types had been the order of the day for more than 300 years, the 1625 to 1925 era, in the Pennsylvania hearth. One factor that determined their appearance was the availability of forest resources from the mid-seventeenth century to late in the 1800s. Forests commonly yielded abundant, large trees until the early 1800s. By then, first-growth and some second-growth forests had disappeared and conditions approximated environments that existed in much of post 1300 Europe. Barn builders and farmers persevered with erecting barns and a plethora of types and subtypes continued to be built throughout much or all of the nineteenth century. The first half of that century has been called the golden age of agriculture in Pennsylvania and saw a multitude of different kinds of barns being erected.

# Chapter Three: Harvesting Timbers from North American Forests

No treatise on North American barns can be complete without some discussion of trees and wood products in the seventeenth, eighteenth, and nineteenth centuries. The more we know about forest resources, wood species, and the conditions under which they grew, the more we can understand about how barn construction decisions were made. Trees supplied a great deal of material for barns, even those of stone or brick.

According to Stevenson Fletcher's classic book, *Pennsylvania Agriculture and Country Life, 1640–1840*, it has been estimated that when the first Europeans settled in Pennsylvania, ninety-eight percent of the land—28.8 million acres—was covered with thickly forested land. Virgin or old-growth forests averaged about 17,500 board feet of lumber and ten cords of wood per acre. A board foot of timber is 12 inches by 12 inches by 1 inch thick. Fletcher quotes Swedish explorer John D. Schoepf in his travels between 1783 and 1784: "The forest is so thick that the tree trunks almost touch, by their height and their matted branches making a dimness, cold and fearful, even at noon on the clearest day. All beneath is grown up in green and impenetrable bush. Everywhere lie fallen trees or those half fallen, despite of their weight not reaching the ground. Thousands of rotten and rotting trunks cover the ground, making every step uncertain; and between lies a fat bed of the richest mould that sucks up like a sponge all the moisture . . . One can with difficulty penetrate this growth even a little way" (Fletcher 1950).

It seems likely that the foregoing description applied to most of the state's original forested areas. We can know with certainty that Schoepf was talking about conditions that settlers in many areas of Pennsylvania knew first hand.

A 2001 science symposium in Canada found these features that help define these forests:

Trees in primeval Pennsylvania forests attained diameters of 3 feet or more. Several patches of old-growth forests remain in the western half of the state.

- Uneven or multi-aged stand structure predominates or several identifiable age companions are seen.

Cook Forest in northwestern Pennsylvania has many substantial white pines, many of them more than 3 feet in diameter. Pine forests supplied boards for exterior siding and mow-stead walls.

Matson Woods in Northampton County has dozens of white oak trees more than 2 feet in diameter. One oak is nearly 4½ feet across near its base. Oak supplied countless timbers for the building of area barns.

- Average age of dominant species approaching half the maximum longevity for species (approximately 150+ years) for most shade-tolerant trees. Some old trees are near their maximum longevity (300+ years).
- Standing dead and dying trees are in various stages of decay.
- Fallen, coarse, woody debris on the forest floor is common.
- Natural regeneration of dominant tree species is present within canopy gaps or on decaying logs.

I have explored many areas of the hearth and it is clear that no remaining forest even remotely matches the description set forth at the Canadian symposium. Only Matson Woods, a five-acre tract in northern Bethlehem Township in Northampton County, likely a second-growth stand, approximates these conditions. Oak dominates the forest canopy with perhaps three to four dozen trees that range in diameter at breast height from 2½ to 4 feet. One massive 200-plus-year-old white oak tree is close to 4½ feet across the trunk near its bottom end. Trees in this wooded area likely average more than 200 years old.

Other geographic sections in Pennsylvania but outside the hearth area still maintain somewhat significant virgin tree conditions. Several examples occur in the northwest area such as Cook State Forest, Tionesta Creek, and Hearts Content, among others. A visit here in the summer of 2004 revealed the presence of many trees 2 to 3 feet in diameter. These areas are mostly home to softwood trees such as white pine and hemlock,

The white oak at London Grove Meeting in Chester County is the largest oak in the state at 7 feet in diameter. This is a field-grown specimen as opposed to the forest-grown trees of pre-1700 forests.

and some hardwoods. This is different from most areas in the hearth where significant numbers of oak trees and other hardwoods once grew. Oak was the predominant species used in barns of all types in the hearth region.

Trees in Matson Woods are forest-grown as opposed to field-grown. Field-grown trees normally have no other trees near them, allowing them to reach for sunlight in most directions. Therefore, these trees favor lateral rather than vertical growth. Forest-grown trees are normally crowded, and branches seek the available sunlight by stretching vertically. Thus, field-grown trees typically have a larger girth than forest-grown trees. This is important to note, since the vast majority of large trees the colonists saw were forest-grown trees.

One example of a field-grown tree in the hearth area is the massive 7-foot-diameter white oak at the London Groves Friends Meeting in Chester County. This is currently the largest oak of any species in the entire state. The tree is in excellent health and it may be approaching 350 years in age. Another example is the well-known Sacred Oak in Oley Township in Berks County. This nearly 6-foot-diameter chinquapin oak has a long-standing connection to Native Americans, and copious feedings have helped it survive.

## Plentiful Oaks

Oak trees were particularly plentiful, though other wood species were certainly used, such as chestnut, tulip poplar, and probably walnut and pine. Settlement-era barn builders probably preferred oak because they were so familiar with the species, and the available timber sizes would have influenced barn design.

We know the general sizes of oak trees that the first hearth settlers saw. In the *Textbook of Dendrology*, William Harlow, PhD, and Ellwood S. Harrar, PhD, state that "White oak at its best is a large tree 80 to 100 feet high and 3 to 4 feet in diameter. Under forest conditions, white oak develops a tall straight trunk with a small crown" (Harlow and Harrar 1969). Not all colonial-era oak forests had trees of these dimensions, but many probably approximated these sizes.

According to Professor Kevin Nixon at Cornell University, a North American oak expert, the genus *Quercus* is one of the most important groups of woody plants in many regions of the Northern Hemisphere. Oaks dominate various temperate, subtropical, and tropical forests and are also a major component of several chaparral and scrub vegetation types.

Another reason we know that oak was in great supply to the colonists comes from an unexpected source called indentures. These legal documents recorded conveyances of land in the eighteenth and nineteenth centuries from a seller to a buyer that include descriptions of land boundaries. Where sides met were property markers, such as stones and often certain tree species. These markers were also called witness trees and consisted of oak trees, but occasionally hickory and chestnut. Some of these marker trees survive, such as a massive oak tree at the Strasser property in Windsor Township in Berks County that was warranted in the mid-eighteenth century.

Perhaps one of the best illustrations of the supremacy of oak in the northern hemisphere is seen in an arcane publication that Paul Friedrich wrote in 1970, *Proto-Indo-European Trees*. This book explores the earliest

This large white oak, at least 250 years old, in Windsor Township, Berks County, was likely a witness tree, used to mark the corner of land surveys. Such surviving trees in the Pennsylvania hearth are uncommon.

The nearly 5-foot-wide oak in Upper Milford in Lehigh County is a size rarely seen as a forest-grown specimen in the state. First- and second-generation builders in the hearth saw similar-size trees.

known European names that were assigned to eighteen categories of trees. They include birch, conifers, juniper-cedar, aspen-poplar, willow, apple, maple, alder, hazel, nut tree, elm, linden, ash, hornbeam, beech, cherry, yew, and oak. The earliest usage of certain tree names goes back more than 3,000 years. The section on the oak species is by far the longest. Friedrich cites three oak names— ayg tree, perk tree, and dorw tree—used thousands of years ago. Even the acorn was used by the pan-Celtic Druids who would consume acorns so "as to acquire powers of divination." Many other details are included about how the oak species played what may be called a pivotal role in the lives of ancient peoples.

Yet another indicator of the absolute domination and highest regard people had for oak, this time in England, is the book published in 1790 by antiquarian and archaeologist Major Hayman Rooke, *Descriptions and Sketches of Some Remarkable Oaks in the Park at Welbeck in the County of Nottingham*. He writes, "The venerable and majestic oak seems to claim a superiority over all other trees. It was styled by the ancients Jovis Arbor; and the Celtic statue of Jupiter was a tall oak." He states that the ancient Druids "esteemed the mistletoe of that tree above that of all the others." Keep in mind that these are direct statements from a native-born Britisher from more than 225 years ago.

If any doubt persists that oak reigned supreme as the tree species of choice in the eighteenth century and the first several decades of the nineteenth century, it may be removed by additional reading of Harlow and Harrar: "The oak group with its many species is the most important aggregation of hardwoods found on the North American continent, if not in the entire northern Hemisphere. In fact, the central and southern hardwood forests of the United States may be thought of as oak forests, with the other broad-leaved trees playing a secondary role. The oaks furnish more native timber annually than any other related group of broad-leaved trees."

They continue, "The sturdy qualities and appearance of many of the oaks, together with their longevity in comparison with other hardwoods, have made them from ancient times the objects of admiration and worship among the early people of the Old World" (Harlow and Harrar 1969). Thus, the use of oak as a primary building material on the two continents formed a continuum of well over a millennium.

The earliest Pennyslvania builders knew that they had an exceptional resource. William Penn advised settlers to "leave one acre in every ten in woodlot and to preserve oak trees and mulberries wherever they were found" (Neccai 2012). They took advantage of its relatively straight grain, especially the arrow-straight texture of the red oak species. Perhaps surprisingly, the various species of red oak were something new to first- and second-generation barn builders in the New World. Only three oak species grew in Europe, and they were all of the white variety.

## Sourcing Timbers

Under virgin forest conditions, timber framers likely depended on tree sources close to the construction site. In Katherine Carte Engels's book *Religion and Profit – Moravians in Early America*, we learn of the early transport of timbers from a sawmill at Gnaddenhutten, which today is Lehighton in Carbon County (a few miles north of the hearth). Starting in 1747, lumber was sent down the Lehigh River to the Moravian settlement at Bethlehem. In a direct line, the two places are twenty miles apart but the "river distance" was about thirty miles. These timbers undoubtedly originated from old-growth forests. However, Moravians likely depended on other timber resources much closer to their settlement.

As the sources of local forest trees dwindled, barn builders must have relied on trees farther from barn building sites. Denuded forested lands later grew a secondary tree cover that took fifty to seventy-five years to grow large enough to be used for second-generation barns. The wood fibers were lower quality than those of virgin-growth trees. When first-growth wood is compared to second- and especially third-growth sources, the differences are easily discernible.

Even a cursory examination of timbers of second-generation barns reveals them to be somewhat curvy or wavy, though they were adequate for strength and support.

A barn's age can't be determined solely by the size of its timbers. The best way to distinguish first- and second-growth timbers, and perhaps their general locales, is from the science of tree-ring dating known as dendro-chronology. This is a means of dating beams from timber-framed buildings—that is, when timbers were felled in forests from where they originated. Thousands of structures have been dated in the east since the late 1970s. To obtain some insight into second-generation buildings, a recent (March 2010) dendro-dating project involved a townhouse in Philadelphia. Five oak cores were taken and the data indicated five dates of tree felling in 1809. Usually buildings were erected one or perhaps two years after trees were felled. The Philadelphia building was erected in 1810. Bill Callahan, a dendro-chronologist who conducted the study, said ". . . there emerged evidence of an intrinsic construction period for the studied section of the Daniel Charpentier House. The absolutely dated, bark-edged oak samples indicate a construction phase for this building some time during the late autumn months of the year 1809, or (most likely) during 1810" (Callahan 2010).

The striking aspect of these findings is that they suggested the timbers had originated thirty to fifty miles west of the site, possibly in Chester County or Lancaster County. Other examples are known of late eighteenth- and nineteenth-century buildings made from timber sources many dozens or even hundreds of miles away from building sites. This essentially proves that local timber sources were not always adequate during this period.

In the case of the 1809 felled timbers, the number of annual growth rings ranged from sixty-one years to seventy-nine years. Chester County and Lancaster County were initially settled about 1700 and many first-growth forests were cleared between 1700 and 1775. These lands harbored the second-generation forests that may have provided those timbers, which originated from trees that germinated from about 1730 to 1750. The dendro-dates of the generated cores from the Philadelphia townhouse and general range of second growth forests in the two cited counties dovetail particularly well.

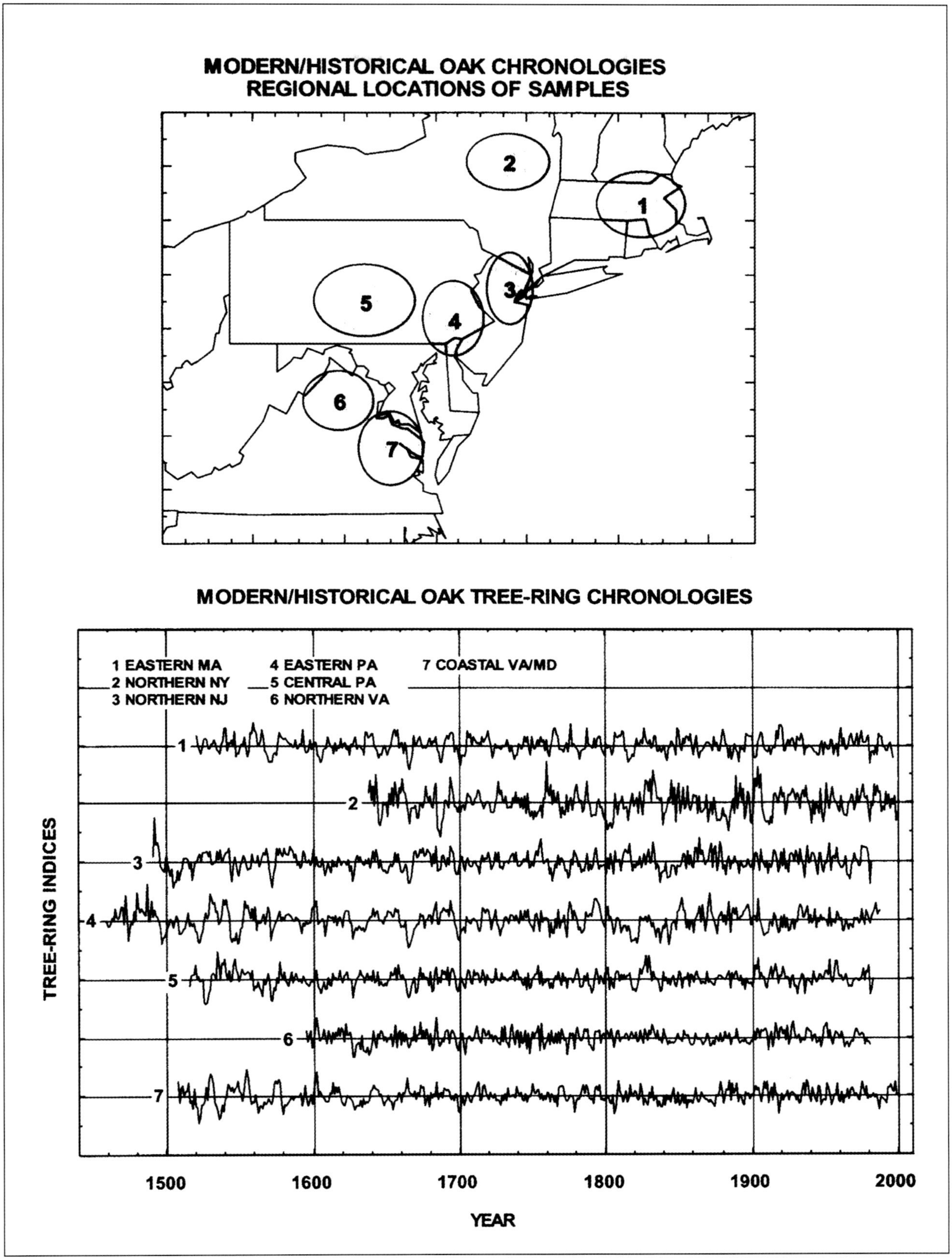


Dendro-date map shows areas of Master Lists created for oak tree chronologies in a few states of the eastern US. (Courtesy of Ed Cook)

A large bank barn in Exeter Township in Berks County provides another interesting case study. The barn consists of two sections—an original circa 1810 section and a later smaller section. The original part consists of small timbers in the bents and roof framing members and basement ceiling joists. The smaller section, constructed a few decades later, has considerably larger basement joists than the original section. What accounts for the difference?

There may have been a local source of timbers for the timber framer to construct the smaller barn section. But more likely the source was from a forest several dozen miles way, possibly from a wooded area north of the Blue Mountain. It is important to remember that transportation increased dramatically after the Industrial Revolution commenced in the 1820s. This permitted large trees or possibly even timbers to be transported many miles. Rivers provided a means of transport from considerable distances, as did new roads (crushed stone), canals, and railroads.

One example of a water route was the Delaware River. Countless timber rafts once rode the spring freshets to markets along the Delaware River where pine and hemlock logs were fashioned into spars and masts for the lordly ships of the British Main. Daniel Skinner was among the first loggers to make the trip down the Delaware. In the 1760s he and two others floated an 80-foot log raft from the Catskill Mountains in New York to the Philadelphia area some 200 miles downstream. Shipbuilders were overjoyed with the fresh supply of timber. It is easy to imagine that Skinner supplied many other areas in the hearth area north of the City of Brotherly Love. He was a pioneer whose ingenuity changed the face of a major waterway.

Another interesting barn, a Sweitzer, in Lancaster County, shows evidence of the use of timbers rafted down a river, probably the Susquehanna River. Nailing boards above the rafters have a series of unexplained large-diameter holes. The holes suggest the former presence of thin poles used to tie all the timbers together on the rafts.

Timbers on the Updike Farm hemlock frame barn in Princeton, New Jersey, were dendro-dated to 1891. Ed Cook, a major figure in the dendro-chronology field who worked on the structure, recalled a forest where he once core-dated standing hemlock trees. The data match from the forest and the building "was staggeringly high." Cook was convinced, almost unimaginably, that the hemlock building timbers originated from the Tionesta forested area of northwest Pennsylvania. Straight-line distance between the two places is about 260 miles. The timbers could have been transported from the Tionesta area to Princeton via railroad. Such great distances were likely duplicated innumerable times in the 1800s via railroad, canals, and rivers.

The Lambertville area along the Delaware River in Hunterdon County, New Jersey, once possessed old-growth forests. This area is so close to Bucks County in the hearth area that some deforestation conditions would have occurred there. Sarah A. Gallagher, in her work, *1703–1903: Early History of Lambertville, New Jersey*, lamented:

> The hillsides are shorn of their forests,
>
> Handsome dwellings adorn the plateau.
>
> What'er was romantic or rustic,
>
> There is naught of it left, that I know.

We will never know even a small number of the sources for structural members in so many hearth-area barns. Still, it is likely that timbers used in most first- and second- generation buildings came from the hearth itself. What we do know is that a broad array of challenges and provocations presented themselves to settlement-era craftsmen, who had to experiment with new ways of forming and joining timbers of many different sizes and tree species. If a dozen barns built during the Revolutionary War could be compared to a dozen barns built after 1820 or 1830 (when the Industrial Revolution started), innumerable differences would be seen.

Some of these differences are seen in the tool marks left on timbers, as the tools themselves underwent an evolutionary process. It is fair to say that if a timber joiner of 1780 were to have met a joiner from 1840, a lively exchange would have ensued.

Ninety percent of the barns in this book are constructed mostly of oak timbers. Some of the beams are from first- growth forests while others are from second-generation forests.

# Chapter Four. Ground Barns

The physical appearance of the earliest barns in the hearth area is unknown. But it is reasonably certain that pioneering barns, wherever they occurred, were one-level structures known as Ground barns. That is, they had no full floors below the wagon floors and therefore were not banked, as were Pennsylvania forebay barns and English Lake District barns.

Some Ground barns were built into banks but the interior compartments—wagon bay and two end bays—were built along the contours of sloped land and had no basement.

The Moravian Archives in Bethlehem, Pennsylvania, contain a sketch of what is almost assuredly a Ground barn of frame construction from the eighteenth century. The rendering includes both longitudinal and transverse elements of timber framing including bents.

In the archives is also a large sketch of the Crown Inn property on the south side of the Lehigh River in

Unruh house-barn combination building from the early eighteenth century in the Germantown area of Philadelphia. The structure was removed in the mid-1800s. (Courtesy of Germantown Historical Society)

The Conoy-area barn in western Lancaster County is a unique Ground barn. It has partial posted corners and a Principal Rafter System and was built in the last decade or two of the eighteenth century.

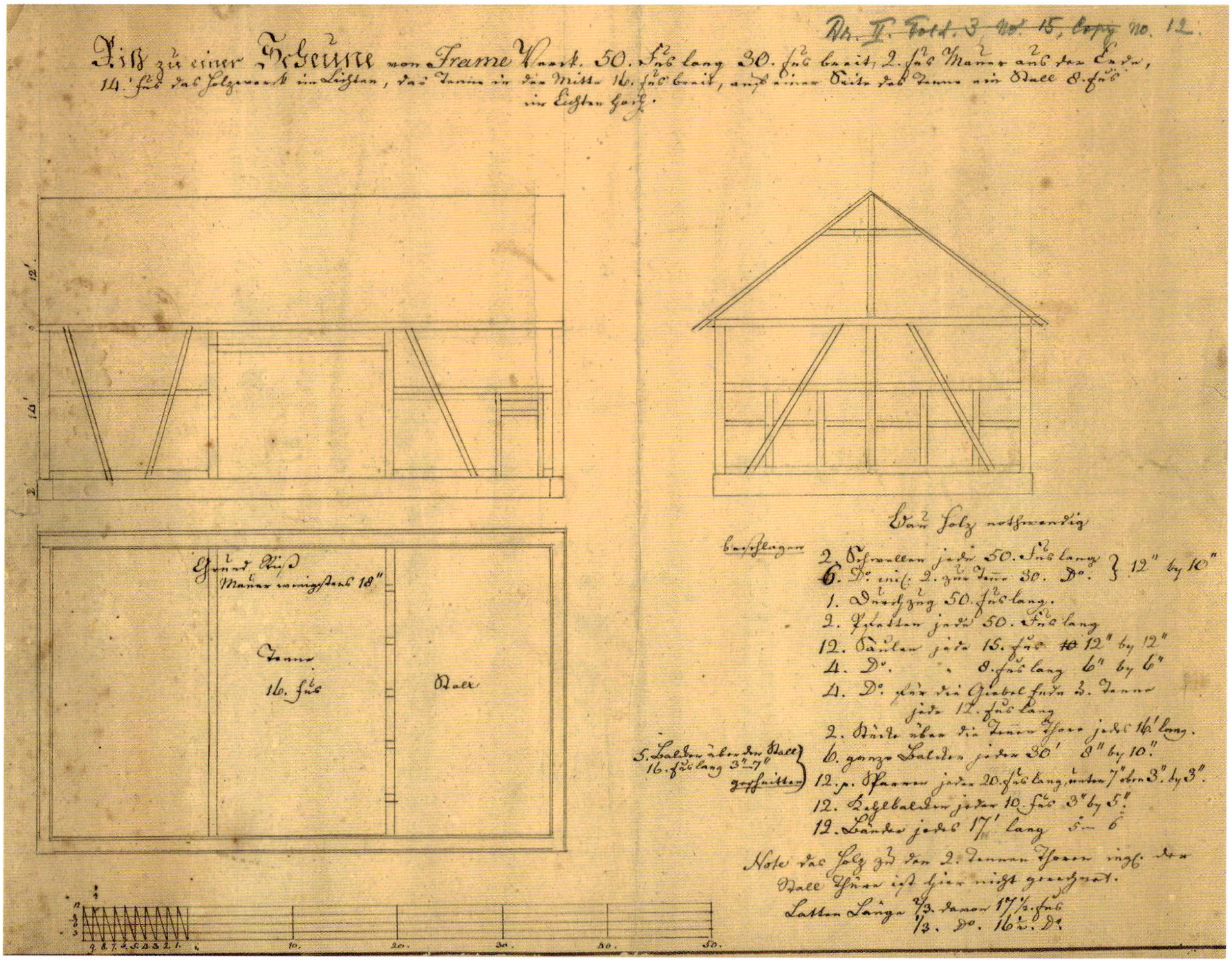

Sketch of probable eighteenth-century frame Ground barn showing various construction elements. (Courtesy of The Moravian Archives in Bethlehem, Pennsylvania)

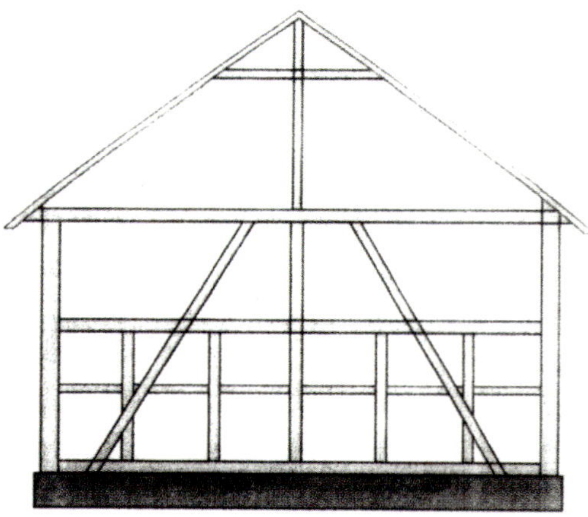

A frame Ground barn includes many of the main framing members of a transverse bent including upper tie, long braces, posts, lower studs, and horizontal girts. No basement is present in this barn form. (Courtesy of The Moravian Archives in Bethlehem, Pennsylvania)

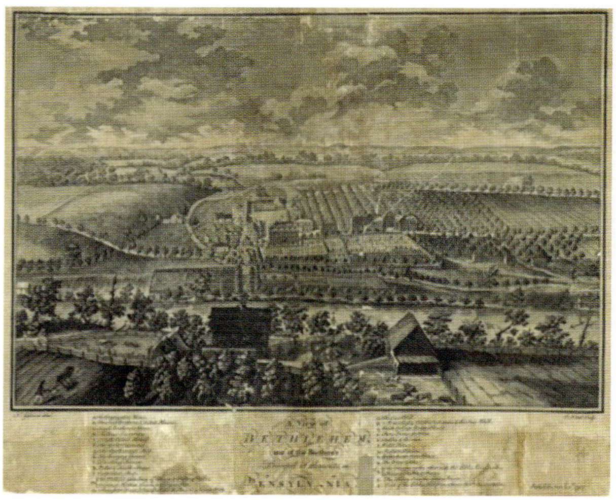

A view of Bethlehem in the 1750s looking north with the Crown Inn south of the Lehigh River. The inn includes a view of a steeply roofed apparent Ground log barn. (Courtesy of The Moravian Archives in Bethlehem, Pennsylvania)

The Heckler stone-to-eaves-level Ground barn in Lower Salford Township in Montgomery County was built in 1761. This is both one of the oldest and most original barns in the entire state.

Bethlehem. This is an illustration of a German property in the 1750s. It includes an off-center-chimney German house and a Ground barn of probable log construction with a steep roof. It includes a side wall wagon entry and what is likely a thatched roof, as was common at that time.

There are small differences in the ways various ethnic groups built their Ground barns. Germans called these barns either *grundscheier* (Ground barn) or *boddamscheier* (bottom barn).

These earliest Ground barns were most likely log construction. A few survive from the third quarter of the eighteenth century. One at the Barnett Heller homestead in Richland Township, in upper Bucks County, was a three-bay double log crib barn that was moved to another location in Bucks County in early 1997. Another resides near Pricetown in Berks County on its original site. This barn may have a carved date in the 1750s. At least two mid-eighteenth-century stone barns are known, and they have steep roofs. The dated 1761 Heckler stone and frame barn in Lower Salford Township in Montgomery County is a good example.

A similar Ground barn circa 1770 resides in Upper Frederick Township just a few miles from the Heckler barn. Many dozens of these barns in strictly German settlements have survived in the hearth area and date from the 1760s well into the second half of the nineteenth century.

English settlers built similar-looking Ground barns. One exception is the well-delineated Lancashire style of Ground barn, which has antecedents in the Lancashire area of England south of the Lake District in the northwest.

No other ethnic-style Ground barns have been identified to date. Other styles were likely built in the first half of the eighteenth century but have long since disappeared from the hearth area. Post 1775 examples may exist but have also not been identified.

The Abiah Taylor stone Ground barn west of West Chester in Chester County has a timber dendro-dated to 1753. It is the oldest barn in Pennsylvania that survives in essentially original condition.

The first treatise on Ground barns was written by Alan Keyser and William Stein (1975). It includes illustrations and photographs of stone, log, and frame Ground barns. Nothing else appeared in the literature on Ground barns until the publication of "Oddities Observed: The Old Ground barns of Oysterdale" in the *Historical Review of Berks County*, Winter 2010–2011 (Huber).

This article involved nine barns in a confined area of eastern Berks County in Oley, Pike, and District Townships. In addition, I wrote about four barns in an article, "One-level Ground barns in Longswamp Township," for the Longswamp Township Historical Society Newsletter in April 2014.

Keyser and Stein stated that they had examined fifty Ground barns. They claimed that, due to "expanded farm practices . . . proliferation of the bank barn, suburban sprawl and the end of the small family barn," the number of Ground barns would be far less than the original numbers that were in the thousands. Since that time, in 1975, I have identified more than 200 Ground barns in fifteen counties, of which more than 175 have been examined and recorded to varying degrees. Many

The Moyer gable forebay frame Ground barn in the Oley Valley, Berks County, was built about 1870. Within just a few miles of this barn are about ten other Ground barns with forebays.

The two-section stone Ground barn in Longswamp, Berks County, is one of several structures with an original two-bay section built circa 1850. The frame addition is circa 1880.

The Pike Township, Berks County, gable forebay barn was built in the late 1870s. Ground barns with stone barnyard walls are quite rare.

facts about these barns have been revealed.

I've chosen to use the term "one-level Ground barn" rather than the term "tri-level Ground barn" that Keyser and Stein promoted. This latter term can only be faithfully applied to a number of late-eighteenth-century and early-nineteenth-century structures. Many Ground barns have no pretense of having three distinct levels among its three bays. The term "one-level Ground barn" can be applied to all of the 200 Ground barns thus far identified in the hearth.

## Description of the Ground Barn Style

Ground barns are usually gable-ended, one-level structures with three bays, though a few four-bay barns exist. Several two-bay Ground barns have also been identified. Three variant styles of barns also fall under the major category of Ground barns.

A three-bay barn, by far the most common, has a central wagon bay and two end bays. In many pre-1810 Ground barns, the two end bays are positioned at different ground levels than the wagon bays and each other. This is the reason that Keyser and Stein offered the phrase – "tri-level Ground barn." In these barns, typically, a few horses occupied the lower level. This side was often at the house end of the barn. The higher end bay stabled a few cows. Above each stable was a mow area where farm crops were stored. Crops were stored from the level of the boards (just above the stable ceiling beams) to the roof peak. Barns typically had more cows than horses, as indicated in estate inventories.

It is interesting to note that the ceiling levels in each of the stable areas vary by 6 to 10 inches or more in a given barn. The horse stable ceiling is higher simply because horses are taller than cows. This differential is consistently seen across cultural lines and has also been seen in hundreds of Holland Dutch-American style three-aisle one-level barns in eastern New York and northern New Jersey.

On eave walls of Ground barns, doors, also of varying heights, lead into stable areas. Even from a distance, Ground barns can often be identified just by the position of the eave wall doors.

The big wagon doors that appear on each eave wall in line with the wagon bays are either batten or frame construction. Original doors are not common. A door with an authentic frame construction survives on an 1870-era barn in northeast Oley Township in eastern Berks County.

Farms with German Ground barns rarely have stone barnyard walls. One exception is off Oysterdale Road in Pike Township. Here the barn is a gable wall forebay example where the positioning of the forebay with its several stable wall doors may have lent itself to a walled barnyard installed adjacent to it.

Crawl spaces 1 to 2 feet high are found below the wagon floor. Above the wagon floor is an *oberdenn* where farm crops were stored. According to Keyser and Stein, unthreshed oats were stored there.

Ground barn footprints are much smaller than the bank barns. A survey of four three-bay Ground barns at German homesteads (or at least in German neighborhoods), built between 1761 and 1791, showed an average length of 60.4 feet, though the lengths varied widely. Curiously, all four barns had a similar width of about 33 feet.

The longest Ground barn documented is outside the hearth and is 75 feet long, whereas no Ground barn in the hearth area is over 70 feet.

*Note: In this chapter through chapter nine, many construction elements are introduced. For explanation of these terms, see chapter ten or the glossary near the back of the book.*

## Roof Structures

The great majority of Ground barns of any ethnic type have Common Rafter Systems. This makes structural sense as most of the barns are small and do not need the high-load-bearing English-based Principal Rafter System or the German-based *liegender Stuhl* Truss System.

Most pre-1850 barns have hewn rafters pegged at their top ends. After about 1850, barns had milled

This Bucks County stone and log Ground barn in the Argus area was built about 1800. It was recently restored to museum-like condition. Its Common Rafter System is shown.

rafters installed while still maintaining pegged upper connections. Late in the nineteenth century, the barns had milled rafters that were butted and nailed at their upper ends rather than pegged.

Only eight barns have been seen that do not have the Common Rafter System. Four of these barns have German truss systems. Three occupy German homesteads, while one, the Stapleton barn in Oley Township in Berks County, may be on an English-settled property.

The 1780s Stapleton Ground barn has one of the steepest rooflines of any barn in the hearth. The barn has a completely original Germanic roof structure along with a rare, swing beam-like construction.

A unique Ground barn of supposed eighteenth-century vintage is in New Hanover Township in Montgomery County. Amazingly, this rare frame construction seems to have originally been a Sweitzer and maintained its German roof and forebay.

Four of the barns have Principal Rafter Systems. Three of them occupy German properties and one is seen on an English homestead just west of West Chester in Chester County. This 1754 structure is the earliest dendro-dated barn in Pennsylvania—the Abiah Taylor Lancashire-style Ground barn. One of these barns, not in the hearth area, is the amazing eighteenth-century Franklin County double log crib barn that includes an astounding seven Principal Rafter pairs. This reflects the fact that the barn is, remarkably, 75 feet long. Ground barns in the hearth area may have attained such lengths. Full two-level Standard forebay barns this size are considered quite large. A photo in the Shoemaker book in the chapter, "The Log Barn" (by Henry Kauffman) illustrates an apparent one-level Ground barn of log construction that has, amazingly, six bays. The barn, with two or even three additions, likely approaches 90 feet long. One of the Ground barns, in Lancaster County in Conoy, is an eighteenth-century structure that has unique partial corner posting. This barn may be the only log Ground barn with a Principal Rafter roof structure. Another barn of unusual brick construction with this roof type, built in 1844 at the Kreibel homestead of the Schwenkfelder sect, is in Hereford Township in Berks County.

One unusual barn, with stone-to-the-peak construction, in West Vincent Township in Chester County, has a unique original roof silhouette. The rear roof has one pitch, while the front roof has two different pitches. This asymmetry is characteristic of up-country posted forebay barns. It is evident that builders used a variety of roof structures 150 to 250 years ago.

## Ground Barn Classification

Ground barns incorporated a wide variety of types and building materials, especially after 1790. They are grouped into two main classes and three minor classes. The principal materials of the first two classes are log and stone.

### Class 1: Classic Ground Barns

These barns have four straight walls without projections or indentations of any kind. To date, classic Ground barns are divided into twenty-five types. This list ignores any distinction between German and English structures. The first four categories of materials usage constitute a progression from what was likely the earliest Ground barn type to the latest type, or log to stone to frame to brick. Eras of construction certainly overlap in each of the four types. For example, some stone barns were built either before or during the eras of log barn constructions. The apparent number of barns in each type is included.

Several of the barns in the first two types are combination barns, that is, they combine two materials

The dated 1844 Kreibel brick Ground barn is just off Route 100 in Hereford, Berks County. It may be one of the rarest brick Ground barns with a Principal Rafter System.

The 1791 Caspar Maul stone-to-peak Ground barn is one of the substantial barns of its type in the hearth. It has a German roof structure.

in their construction. The two other types are pure constructed types—only frame or only brick. Specific numbers of barns in most types and sub-types are mentioned at the end of the chapter. When types and sub-types have just one barn in each category, then they are designated as such.

### Class 2: Ground Barns with Forebays

This forebay style has a cantilevered section that extends beyond an end wall or an eave wall. Barns with this characteristic have just one wall projection. No original Ground barn has been identified with projections on two or more walls. However, one two-section barn in Bethel Township in Lebanon County has two end wall forebays, one in each section.

This frame gable wall forebay barn in Rapho Township, Lancaster County, is one of several barns of this type within a few miles of each other.

It should be emphasized that Ground barns with forebays are not part of the population of Pennsylvania German–style bank barns, all of which have forebays. Ground barns are strictly one-level unbanked structures, while all Pennsylvania barns are two-level banked structures with forebays.

Forebays protect the stable wall doors and increase storage space in the mow. The earliest Ground barn with a forebay is the stone structure on Route 322 in East Brandywine Township in Chester County dated 1814. The forebay is at the gable end wall. It is likely that few, if any, Ground barns of the eighteenth century had a forebay. It also seems probable that some homestead barns incorporated forebays inspired by Standard barns in the hearth area.

The Heller-Moyer double log crib Ground barn in Richland Township, Bucks County, was one of the oldest intact Ground barns of its type in the hearth until the late 1900s. It was dismantled in late 1996.

**Ground Barns by County**

Ground barns have been found in all eleven counties in the hearth area. The following list (right) reflects the extent to which the counties were surveyed—Lehigh and Berks Counties more thoroughly than others. When the less-traveled counties are studied more closely, the number of Ground barns will increase dramatically. More than 150 Ground barns have been included in the list.

Berks County - thirty-eight
Bucks County - fourteen
Chester / Delaware Counties - two each
Lancaster County - eight
Lebanon County - six
Lehigh County - fifty-six
Montgomery County - nineteen
Northampton County - ten
Philadelphia / Dauphin Counties - one each

## Seven Outstanding Ground Barns

The following examples of Ground barns represent different materials, dimensions, and eras. They offer a good general sense of the diversity of these barns.

### Heller-Moyer Double Log Crib Ground Barn, Bucks County

The Barnet Heller barn, formerly in Richland Township, was first illustrated in the Keyser-Stein 1975 article, "The Pennsylvania German Tri-level Ground barn." A floor plan of this three-bay double log crib barn along with a view of a pig door in the cow stable of the barn is provided. In the 1994 Harry C. Adams book, *The Direct Tax of 1798*, a photo shows vertical exterior siding plus a one-bay frame addition. In the 1798 tax list, the barn is described as a 40-foot by 22-foot round log barn. The actual width is 24 feet. Side wall height is 13 feet, 3 inches.

The Heller-Moyer log barn had an original log wall door. The logs were left in the round.

This barn was recorded in December 1996 at the original homestead site. A few weeks later the barn was dismantled and moved to a new site in Spinnerstown about four miles away. It was re-erected within a year and is currently well cared for.

Roof pitch was steep in keeping with late eighteenth-century standards. Roof structure was a Common Rafter type composed of fourteen pairs of rafters. All rafters (hewn) were cogged (tenoned) at their bottom ends to receive slotted wall plates. One roof slope had two wind braces.

Each log crib was 14 feet wide (eave wall side) while the wagon bay was 11½ feet wide. The northeast end bay contained a feed entry at the south side and a three-pen horse stable at the north side.

The stable was entered through a door on the north eave wall. A longitudinal staked manger was located where the feed entry and the stable met. The six-pen cow stable was at the southwest side where corner doors at each eave wall entered into this stable. A transverse staked manger was located where the crib and the wagon bay met.

Importantly, each stable area had different ceiling heights. The horse area ceiling was 6½ feet high, while the cow area ceiling was 5 feet, 10 inches high in response to the animals' height differences. This stable ceiling height differential is almost universal in one-level Ground barns in the hearth area.

Each log wall of each crib consisted of thirteen logs. The round logs were about 12 inches in diameter. The corners had inverted V-notching where the sides of the V were quite steep. This distinctive notching is called "tear-drop." The logs appeared to be white oak. Chinking (in-fill) existed on all four walls of one log crib.

Above the stable areas in the end bays were mow areas where farm crops were stored. These areas extended from the stable ceilings to the peak.

The wagon bay was commonly used as a threshing area in vernacular barns until the late nineteenth century. The floor planks were 2 inches thick, likely oak, and up to 1 foot wide. The Heller log barn likely dates to the last quarter of the eighteenth century. Every feature essentially agrees with this estimate.

### Heckler Stone-to-Eaves-Level Ground Barn in Lower Salford Township, Montgomery County

The Heckler barn off Quarry Road is one of the most outstanding barns in all the hearth area, if not the state. It may be the earliest Ground barn in the hearth that is essentially still in original condition. A construction date of 1761 is carved on its overhead wagon bay entrance lintel. Recently, a new wood shake roof was installed. This venerable old structure is well cared for and open for visits at Heckler Plains Township Park.

The barn has stone to the eaves level on its gable ends and measures 56 feet long by 34 feet wide. The roof slope is quite steep. All exterior walls have splayed loopholes that are each eighteen inches in height. The

The Heckler Ground barn has a different treatment of doors at its end walls. The wall at the cow end has two openings, while the horse end has just one opening.

ground slopes down toward the house or southwest side of the barn. This lower level contained the cow stable, in contrast to other early Ground barns where the lower level was almost always used for stabling horses. The house end wall has two doors, one for a feeding alley and the other for cows. One eave wall has two doors at either side of the wagon doors, each leading to a stable. The far end wall has a single door leading to a feeding alley.

The roof structure is typical of early German roofs and consists of two classic truncated Principal Rafter truss pairs and seventeen Common Rafter pairs. All original timbers are present. Curiously, the longitudinal *oberdenn*-level beam cracked and was later augmented by another beam placed to its immediate side.

The wagon bay is 16 feet wide, the horse end bay is 15½ feet wide, and the house side end bay is 20½ feet wide. The house side bay is wider to accommodate the cow stable. The cow stable is a few feet lower than the horse stable. No original staked mangers or manger troughs remain.

### Stone-to-Eaves-Level Ground Barn in Upper Frederick, Montgomery County

This circa 1770 structure is possibly the second earliest surviving German stone Ground barn in the hearth. The barn is down a long farm lane and when the barn first appears on the landscape, the steep roof is particularly striking.

The barn measures 57 feet long by 33 feet, 8 inches wide. The eave wall height is a rather tall 15 feet, 9

The stone Ground barn in Upper Frederick, Montgomery County, was built circa 1770. The lower stable areas are gone, but the barn has the most interesting German roof structure of any barn.

inches. The barn, except for much of its upper roof structure, has experienced many alterations, particularly on both eave walls where the wagon doors were located.

The original basement areas for stabling farm stock are almost completely unrecognizable. Still, the widths of the three bays are discernible. The wagon bay was 14 feet wide. North end bay was 18 feet, 3 inches wide, and the opposite end bay was 24 feet, 9 inches wide. This last bay is wide by Ground barn standards.

The only original section that remains essentially intact is the roof support system. This is a *liegender Dachstuhl* structure, the German dialect term for lying or inclined chair. Four Principal Rafter trusses with Common Rafters constitute the entire roof support. All the trusses were originally classic style. Many of the timbers of each truss except the truncated Principal Rafters have been removed in the past decades. All Principals are milled. Each Principal has a decided taper—about 8 inches wide at their bottoms and 13 inches at their top ends. Purlin plates are set square (not canted or angled) into notches at the top ends of the Principals. The sections of the "secondary" or lower plates are set end to end (not staggered) and appear about 3 feet below the primary purlins. Three secondary purlins are positioned on each roof slope.

Common rafter arrangement (among Principals) is three, three, four, two, two. Twenty Common Rafters are present in the entire barn. All commons are milled and are cogged (transverse tenons) at their bottom ends to receive the slotted wall plates. Such cogged rafter bottom ends and attendant slotted plates normally indicate an eighteenth-century construction date.

The *oberdenn* beams consist of two longitudinal barn-length beams or one on top of the other, separated by 8 feet vertically. This construction is unique on any German roof. Most of the two beams are gone but remnant sections and notches at the bottoms of three upper ties that received the lower beam indicate that both longitudinal beams stretched the full length of the barn. The lower beam was in its normal position. The upper beam was placed between the upper collar beams and the lower straining beam. In addition, four transverse beams originally rested on the lower longitudinal beam. Now just remnant stubs remain on top of the upper walls plates.

Wedge-shaped false rafters, or *Aufschiebling*, attached to the Common Rafters over-sail the eave walls, apparently to protect the walls from the weather. These "rafters" are 3½ feet long and project over the wall by about a foot. They create an attractive kick in the roof and even casual passersby can identify a German roof by observing this distinctive characteristic. Eighteenth-century German houses and barns often have false rafters.

This stone Ground barn, despite its number of alterations, is still a gem. Its rare roof structure and unique *oberdenn* beam system and steep roof slope all attest to its venerable position among Ground barns in Pennsylvania.

## Stapleton Ground Barn, Berks County

The Stapleton three-bay Ground barn is down a long farm lane in Oley Township just to the west of Manatawny

The Stapleton barn was a drive-through structure in which loaded hay wagons entered through one eave wall door and exited through the opposite eave wall.

The Stapleton barn wagon floor was widened with a 4-foot-wide alcove where grain was threshed as the animals walked in a circle.

Creek. This is one of the state's most original eighteenth-century Ground barns. It has several outstanding aspects including a unique "swing beam" construction.

Stone-to-peak, its eave walls are 68 feet, 3 inches long, making it the longest Ground barn in the hearth. The end wall is 33 feet, 2 inches wide. Side wall height is 13½ feet.

The front wall has a pent roof over two cattle doors. The roof covering consists of rare red tiles. On the fascia board is a date of 1782 that is not likely original to the barn. Sliding wagon doors are not original.

Both end walls have vertical splayed loopholes in an arrangement of one over two over three. The south end wall has a single stable level door while the north end wall, in a rare treatment, has three doors. The rear eave wall has a wagon door entry but the door is not original. Each eave wall also has loopholes, as well as kicks at their bottom ends indicating the presence of false rafters.

The *liegender Dachstuhl* roof structure consists of four classic trusses. Each truss retains its original timbers, making this barn one of the few eighteenth-century structures with an entirely authentic roof support. The Principals are milled, but the Common Rafters, twenty-one pairs in all, are hewn and follow a three-three-three-four-four sequence; there are twenty-one Common Rafter pairs.

Curiously for a structure of this size, the Stapleton barn does not have an *oberdenn* longitudinal beam, which lends stability. However, the barn has its full complement of regular *oberdenn* beams.

The wagon bay is 12 feet, 10 inches wide. The north end bay that includes the horse stable is 29 feet, 10 inches wide—possibly the widest bay in any Ground barn. The south end bay with the cow stable is a wide 22 feet, 5 inches.

The width of the end bay with the cow stable includes the 4-foot-wide area under a cantilevered area continuous with the wagon bay. This unusual area is a result of a unique "swing beam."

Swing beam barns principally occurred in the Schoharie and Mohawk River Valleys in New York State, west-central New Jersey, southeast Bucks County, and sporadically elsewhere in the hearth area. The barns were side wall entrance structures, either one-level Ground barns or forebay bank barns. The defining element is a large (up to 2 feet thick) transverse timber, part of a bent that flanks the wagon bay. The beam has no posts, studs, or braces below it, creating an open area below it. Apparently horses or oxen were able to walk in a circle to thresh the grain. Another theory is that the swing beam allowed the horses to turn around and exit the barn after they were unhitched from wagons.

In the Stapleton barn, a normal eave-wall-to-eave-wall swing beam was not included in the construction in the wagon bay area; instead, an alternative and novel structural idea was implemented. The series of ceiling joists in the cow stable area was extended 4 feet beyond the mow-stead wall. These joists supported a transverse beam (a part of the framing unit next to the wagon bay) that was about 7 feet above the wagon floor. No other barn in the northeast or Pennsylvania has this swing beam style. This may be one of the earliest incorporations of a swing beam in a Pennsylvania barn.

More than half of the north mow-stead wall that abuts the wagon bay is original, and the south wall adjacent to the cantilever is fully original, as are a few doors. There are some unusual carvings on the boards of the north wall.

A cryptic carving on one of the original mow-stead walls in the Stapleton barn. The carving is unique in the hearth area.

Eighteenth-century wrought nails secure the tulipwood boards. Except at the ends, the wagon floor consists of all original floor planks. Planks are pegged and up to 16 inches wide.

The horse stable area is immense for a ground-level structure. The floor is 4½ feet below the wagon floor. This drop-off may be the greatest in any tri-level Ground barn in the state. The middle end wall door (one of three) leads to a rather wide feeding alley. All original staked mangers and manger troughs in the stable have been removed.

The cow stable floor is less than two feet below the wagon floor. A strangely positioned longitudinal stone wall is 8 feet, 3 inches from the rear eave wall. Stretching the full width of the end bay, its function is unknown.

The barn was probably built in 1780 to 1790 era, judging by its steep roof and the use of both wrought nails and eighteenth-century hardware. It has all the hallmarks of German construction, yet the homestead has a supposed English settlement history and the Stapleton family originated from England. The barn is one of the most important buildings of its kind in the entire hearth area—a gem.

The Kline stone and log Ground barn is one of the better-known barns of its type. The original section has three bays circa 1810, and the frame addition was built in the late nineteenth century.

### Kline Stone and Log Ground Barn, Berks County

The Kline family owned this property, south of Bowers, in the later half of the twentieth century. First seen in May 1991, the barn appears on the first page of the 1975 Keyser-Stein article. Despite its small size, it is a classic example of the barn type and was likely built circa 1810.

This is a good example of a combination stone and log structure. The higher side of the front wall has a 6-foot-tall stone section while the lower side has a 9-foot-tall stone section. Above the stone areas, logs are covered with vertical siding at the left side and original horizontal oak siding on the right. True to form, the cow side is on the higher side. The horse side opposite is a few feet lower than the cow side. The front wall of each stable area has an entry door.

The interior has a Common Rafter system. The corners of each log wall in each log crib feature inverted V-notching. The sides of the V are moderately sloped and provide a good indication of its general construction era—in the first or second decade of the nineteenth century. There are several long, wide notches in some of the crib logs, suggesting that the logs may have come from a former log church nearby and indicated the window placement.

### Conoy Township Double Log Crib Barn with Partial Corner Posting, Lancaster County

One particularly intriguing log barn of any type in the hearth is a double log crib Ground barn near the Susquehanna River in western Lancaster County. It is unusual that the log cribs seem to have been built at different times. The crib closer to the stone German homestead house has logs with flush corners—that is, they do not extend past the point where they join each other. The logs on the other crib are crowned—they extend about 6 inches beyond the point where they join.

The Conoy barn features unique partial corner posting on the house-side crib; the four corners are posted at the bottom 5 feet where the horizontal logs are tenoned into corner posts. Above these posts, the logs are notched in the normal manner. A few logs have been replaced in the past few decades.

The three-bay barn is 52 feet long and 25 feet wide. The logs of both cribs are not covered with vertical siding, but are exposed. On the rear eave wall is a door, possibly original.

A rare Principal Rafter System occupies the house-side end bay. Here the structure is composed of two Principal Rafter pairs. The Common Rafters are hewn. In addition, over the wagon bay and the far end bay are hewn Common Rafters.

The loading openings in the log walls of the two mows facing the central wagon bay suggest that the barn was built in two stages. These openings were created for farm crops to be unloaded from wagon bays into the mows. While most such openings have straight vertical sides, the house-side log mow opening has angled sides, making the opening trapezoidal in form. In the other log mow, the loading opening has vertical sides. The stable area of the house-side log crib has remnant staked mangers and a thick plank manger trough, possibly original.

This is one of the most remarkable log Ground barns remaining in Pennsylvania. The house-side log crib was likely built in the eighteenth century. Whether more of these barns with both the rare partial corner posting and a Principal Rafter System were constructed in the 1700s is unknown. The barn is in excellent condition.

On the end wall of the Conoy barn, the outer ends of the stable ceiling joists are exposed on the exterior. The joists are seen just above the corner posts and appear as a series of beams in a straight line.

The unique Conoy-area barn in Lancaster County is corner posted. This rare construction feature may have been more prevalent in the eighteenth century.

### Neumoyer Stone-to-Peak Ground Barn, Lehigh County

The Neumoyer barn in southern Lower Macungie Township was likely built between 1820 and 1840. It has a frame one-bay addition and a small frame lean-to shed. Exterior dimensions of the Ground Barn section are 36 feet, 3 inches long and 24 feet wide.

The front wall consists of original centered board and batten wagon doors. Inside, where the door halves meet, is a rare vertical oak bar release. Original boards above the wagon doors are punctuated by six "pigeon holes," an uncommon feature. A door at either side leads into a stable area at each end bay. The original split door with a *spriggel* (bar) was on the horse side.

The Lower Macungie Township stone-to-peak Ground barn was erected circa 1830. This small barn occupied a small farm. It is one of only two fully stone three-bay barns in the township.

The barn has a milled Common Rafter system of fourteen rafter pairs. Since it is small, no purlin plates were necessary.

Three-bay construction defines the barn – a wagon bay and two end bays. Wagon bay is 11½ feet wide. Horse side end bay is 10 feet, 9 inches wide. Cow side end bay is 11 feet, 2 inches. Some original *oberdenn* beams are intact over the wagon bay.

The east mow-stead wall is mostly original with pine boards secured with cut nails. The west wall no longer exists. The two bents that flank the wagon bay have the normal upper tie beams and upper- and lower-level partial barn-width girts and built-in ladders. The ceiling joists of the horse stable are 7 feet, 1 inch above the floor while the joists in the cow stable are 6 feet, 4 inches above the floor.

## Classifying Ground Barns

The following select list of about 145 Ground barns is a general overview of what barns in the hearth area looked like across a 125-year period—until 1850 or even later. Most are in the hearth area. Materials and construction techniques changed through the decades. Each type was the farmers' response to specific conditions. The builders were not locked into a mindset that produced only a few variations of the basic Ground barn structural layout. It is likely that an additional 200 Ground barns survive. Each category under class 1 is the principal building.

## Class* 1: Classic Ground Barn

### Log

| | |
|---|---|
| Type A: | Double log crib Ground barn |
| Sub-type A: | Double log crib Ground barn with raised roof - one |
| Sub-type B: | Double log crib Ground barn with one-bay frame addition - one |
| Type B: | Half stone and half log (each end bay) - a few barns |
| Type C: | Log (one end bay) and frame (one end bay) - one |
| Type D: | Single log crib - one |

### Stone

| | |
|---|---|
| Type E: | Stone to peak (both end bays) - a few barns |
| Type F: | Stone to within 5 feet of peak, both end bays - one |
| Type G: | Stone to peak (one end wall) and stone to eaves level (other end wall) - one |
| Type H: | Stone to eaves level (both end bays) - a few barns |
| Type I: | Half stone and half frame (both end bays) - eight |
| Type J: | Stone (one end bay) and frame (other end bay) - one |
| Type K: | Stone bottom half and frame upper half (one end bay) and all frame on other end bay - two |
| Type L: | Stone to peak - barn with a swing beam - one |
| Type M: | Double barns or end-to-end stone to peak barns - one |
| Type N: | Half stone & half frame (one end bay) two-bay barn - two |
| Type O: | Stone (one end bay) two-bay barn - two |
| Type P: | Stone (one end bay) two-bay barn with frame one-bay addition - three |
| Type Q: | Stone section (original section) and stone addition - one |

### Frame

| | |
|---|---|
| Type R: | Frame (both end bays) - forty |
| Type R1: | Frame two-section barn, original section four bays and addition two bays - one |
| Type S: | Frame with one-bay frame addition - three |
| Type T: | Frame four-bay barn with contiguous two wagon bays - one (likely a three-bay original barn with one-bay addition) |
| Type U: | Frame two-bay barn - seven |
| Type V: | Frame two-bay barn with one-bay frame addition - two |
| Type W: | Frame two-bay barn plus addition of two bays of frame - one |

### Brick

| | |
|---|---|
| Type X: | Brick (both end bays) - six |
| Type Y: | Brick with wagon bay as far-end bay - one |

## Class 2: Forebay Ground Barn

| | |
|---|---|
| Type AA: | Gable wall forebay frame barn - eighteen |
| Type BB: | Eave wall forebay frame - eight |
| Type CC: | Log (one end bay) and frame (one end bay) eave forebay - one |
| Type DD: | Double log crib with eave forebay - one |
| Type EE: | Half stone and half log with eave forebay (end bay) two-bay barn with frame one-bay addition - one |
| Type FF: | Brick gable forebay - one |
| Type GG: | Eave wall partial forebay frame barn - one |
| Type HH: | Eave wall "split" forebay frame barn - one |
| Type II: | Stone gable forebay with original stone lean-to dated 1814 - one |

* Each category under Class is the principal building material.

### Class 3: Variant Ground Barns

| | |
|---|---|
| Type JJ: | Variant stone to peak - one |
| Type KK: | Variant frame barn with basement below one end bay - one |
| Type LL: | Variant two-bay stone to peak barn with basement below one end bay - one |

### Class 4: Lancashire Type Ground Barns

| | |
|---|---|
| Type MM: | Lancashire style stone to peak - one |
| Type NN: | Lancashire style stone to peak and heightened with stone - one |
| Type OO: | Lancashire style brick - one |
| Note: | Lancashire barns have no eave wall doors for animals. |

### Class 5: Ground Barn to a Standard Barn Form and Remnant Barns

| | |
|---|---|
| Type PP: | Original two-bay frame barn converted into three-bay Standard barn - 1 |
| Type QQ: | Remnants. One barn is stone and frame plus stone and log or frame section, one barn has its wagon bay plus part of one end bay, and one barn dated 1748 is frame but has a nineteenth-century roof structure and was placed on a basement construction – three |

# Chapter Five.
# Sweitzer Barns

A Benjamin Latrobe painting of 1801 of a Sweitzer in western Lancaster County is the earliest illustration of this barn type in Pennsylvania. The barn has all the exterior features of an eighteenth-century Sweitzer. (Courtesy of Maryland Historical Society)

The two-level forebay barn was the most widely built barn style and is still the most prevalent in the hearth. The Sweitzer configuration hints strongly of medieval-like construction. Its front appendage or forebay conveys an unmistakable "low-to-the-ground feeling." This contrasts with later models of forebay barns with high eave walls. Compare the steep roof and low-slung forebay of the mid-eighteenth-century Isaac Long barn in Lancaster County to the high-walled Standard barns, and they seem to be from different worlds.

Sweitzers share at least ten characteristics:

- They have steep rooflines, especially barns that pre-date 1800.

- Asymmetrical rooflines result from forebays that make the front roof slope several feet longer than the rear roof slope. The forebay ends are usually flush with the end walls of the

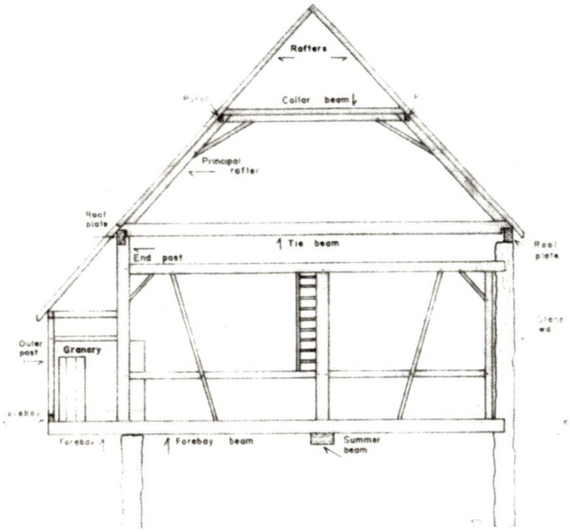


This transverse view of a Sweitzer includes the German roof structure and bent timbers with upper ties, three main posts, built-in ladder, and long braces. At barn front (at left) is the frame forebay with granary. The asymmetrical roof defines this barn style. The basement is stone walled. (Courtesy of Ken Easterday)

barn's main section, but sometimes they are recessed by 2 or 3 feet. An example is the small 1800 Hager Sweitzer in the Harrow area of Nockamixon Township in Bucks County. Sweitzer forebays normally extend 4 to 8 feet.

- The earliest Sweitzers are log, followed by stone, frame, and brick. The earliest stone Sweitzers have vertical slits or splayed loopholes on their end walls. Later Sweitzers have louvered windows, and still later ones have no wall openings on the upper floors. These features are a good general dating tool.

- In most cases, the end walls closer to the house lack basement doors. As we'll see later, these doors are common on Standard barns.

- Forebays project 4 to 8 feet. Sweitzers have three to six stable wall doors, which are usually split into upper and lower halves.

The circa 1800 stone and frame Sweitzer near Womelsdorf, southwest Berks County, retains its original vertical siding on an end wall.

The Philip Hager Sweitzer near Harrow in Bucks County is dated 1800. This is one of few Sweitzers that have forebays with recessed end walls.

- The roof structures employ all three major support types—German, English, and Common Rafter systems. Many Sweitzers have milled forebay rafters while earlier ones have hewn rafters.

- Transverse bents or framing units in non-log barns do not extend into the forebays. Bents make use of posts, ties, girts, braces, and built-in ladders. Typically there are fewer timbers in Sweitzer bents than in other forebay type barns.

- The earliest Sweitzers have mow-stead walls with horizontal oak boards secured with either wrought nails or transitional era nails. Some early barns have pine boards. One hard pine log Sweitzer in Dauphin County built 1800–1810 has, without precedent, mow-stead wallboards (of pine) secured with oak pegs.

- Many of the features in basements of later forebay barns (Standard barns and Extended barns), such as staked mangers and manger troughs, are also seen in Sweitzers. The mangers and troughs on the horse side are longitudinally oriented, while the mangers and troughs on the cow side are transversely placed.

- Many Sweitzers at homesteads had substantial stone barnyard walls. One extreme example is the circa 1790 Moyer barn in western Maiden Creek Township in Berks County, where one section of the wall is 165 feet long and 14 feet high.

The mow-stead wall in the Schaefferstown, Lebanon County, double log crib Sweitzer is original. The wall is pine with wrought nails. This barn, with a carved date of 1796, was dismantled in 2007.

The stone barnyard wall at the Moyer homestead farm in Maiden Creek, Berks County, is 165 feet long and 14 feet high—the largest in the area.

This superbly crafted hard pine double log crib Sweitzer is in Dauphin County. The mow-stead walls are original and, amazingly, the boards are secured with wood pegs.

## Stables and Their Relation to Main Barns

It is important first to explore buildings referred to as stables in the 1798 Federal Direct Tax. It seems likely that these stables housed horses rather than cows. In either case, it is not known how these stables related to areas normally set aside in one-level Ground barns and two-level bank barns such as Sweitzers in the late 1700s.

It seems that between five and ten percent of farms included stables. For example, in Macungie Township in Lehigh County, 9.4 percent of homesteads had stables. Perhaps these stables supplemented the stable areas in regular barns. At some farms, stables were substantially built with stone walls. Most stables, however, were frame. No stable to date has ever been identified as an eighteenth-century building.

In understanding the evolution of barn building traditions in the hearth area, the significance of stables has to be seriously considered.

## Early Sweitzers

A few archival sources may be found on barns of the 1700 to 1750 era. One example is a reference to a barn found in a 1725 inventory of Hans Brubacher in a Chester County probate paper " . . . grain in ye barn." Another is seen in Reverend Benedict Brechebuhl's will in the middle of the first quarter of the eighteenth century concerning his Chester County homestead: " . . . my house and barn with my plantation and land I now liveth on . . . ." Such very early references are reassuring that such farm buildings did indeed exist nearly three centuries ago.

It has long been thought that forebay barns, which came to be known as Sweitzers, first appeared in Lancaster County. This is because the earliest dated Sweitzers are in Lancaster County, and they are found here in greatest concentration. In addition, the 1798 Federal Tax reveals that the longest barns of that era were in Lancaster County. This makes sense, since the middle third of Lancaster County harbored the greatest primeval forest in the region and white oak was plentiful. This indicated to the earliest Mennonite settlers that these lands had the most fertile soil in southeast Pennsylvania.

The term "Switzer barn" was first seen in the *Lancaster Journal* on November 14, 1801, to describe a farm in Donegal Township in Lancaster County: "and [a] large stone Switzer barn." The term "Swisser barn" was also used in the *Lancaster Journal* on August 14, 1802, in reference to a farm in York County: "a large Swisser barn on the premises." Thus we may conclude that the Sweitzer barn was well-established by the turn of the nineteenth century.

Other terms such as "Sweisser," "Swiss," "Switzer," "overshoot barn," and "overshot" are seen in advertisements and property records of the first quarter of the nineteenth century. These are all variations of Sweitzer. The term was probably borrowed from Switzerland, where the barns originated.

While we do not know the precise style of the earliest Sweitzer barns, we know that they had a cantilever at their front side walls. This form—which was easily constructed and made efficient use of interior spaces—probably became stylized in the middle third of the eighteenth century. Sweitzers were typically built with three or four bays, but a few two-, five-, six-, and seven-bay Sweitzers exist.

## Materials Used in Eighteenth Century Lancaster County Barns

In the 1798 Federal Tax list, 1,148 barns were counted in eight Lancaster County townships. The following array of twenty categories of materials are listed in the construction of the barns: "log," "round logs," "stone," "stone and log," "stone and frame," "stone and wood," "three-quarters stone-one-quarter wood," "frame," "lower stone–upper log," "first level stone-second log," "first stone–second log," "part stone – part log," "part stone–part round logs," "square log," "limestone," "limestone and frame," "square log and stone," "lower part stone-upper brick," "lower part stone-upper wood," and "lower part stone-upper frame."

The list discloses the fact that while log, stone, and log-stone combinations were the most commonly cited materials, other materials were also used, even brick, although rarely. Perhaps the most interesting citation is "three-quarters stone-one-quarter wood." This designation may mean a crib or end bay that has three walls of stone and one wall of log or perhaps even frame. It is important to realize that barn assessors might have used different terms to describe barns that were built of the same basic materials in the same mode. They may have also used similar terms to describe barns that may have had different materials. The use of certain words and terms must have been, to whatever extent, rather arbitrary.

The Henry Schreiner stone-to-peak Sweitzer northeast of the city of Lancaster, in Lancaster County, was built in the 1820s. This barn is illustrated in the Dornbusch-Heyl book.

This frame Sweitzer in East Petersburg, Lancaster County, was likely built after 1830.

The log Sweitzer near Shartlesville in Berks County was built around 1800 and is in excellent condition.

## Dimensions of Eighteenth Century Lancaster County Barns

The tax list also provides barn dimensions. While most of us would likely think that most eighteenth-century barns were small to medium sized, perhaps 30 to 60 feet long, seven of the eight townships had at least a few barns more than 100 feet long (Lancaster Township had one). It seems that the barns, especially the quite long ones over 70 feet, may have been built in two or three phases. Two townships, Lancaster and Manheim Townships remarkably each had one barn up to 150 feet long. Manheim had eight barns over 100 feet long. Warwick had fifteen barns over 90 feet long. Any barn over 80 feet or so in length is a substantial structure, even by current standards. Lancaster County had 173 barns with this length or greater. It had forty-two barns 100 feet or longer. Lancaster County was an area of large barns.

For sake of comparison with Lancaster County, the following is taken from the 1798 Direct Tax list and refers to barns more than 100 feet long in various counties. They appear in descending order: Dauphin County, twelve barns; Montgomery County, eight barns; Cumberland and York Counties, four barns each; Franklin and Chester Counties, three barns each; Somerset County, two barns; and Bucks and Huntingdon Counties, one barn each. Bedford, Delaware, Fayette, Greene, Luzerne, Lycoming, Miflin, Northampton, and Westmoreland Counties had no barns. Together these eighteen counties had a total of thirty-eight barns over 100 feet in length. Lancaster County, in distinct contrast, had four more 100-foot-plus barns than these counties combined. In addition, far more barns in the 80-to-100-foot length category existed in Lancaster County at 131 barns, than in any of the other separate eighteen counties, which had a combined total of 228 barns in the 80-to-100-foot range.

## Clue to Lancaster County as First Location of Forebay Barns

The average number of barns more than 100 feet long in each county is 2.1, while Lancaster County has twenty-one times that average. This difference is statistically significant.

It seems that Lancaster County had considerably more fertile soil than any of the other counties. Greater fertility translates to a greater need for storage capacities in barns. As expanding agricultural methods and means of increasing crop production was witnessed in the eighteenth century, especially after about 1750, when lime was used in increasing amounts, the original barns were enlarged. This is the fundamental reason why so many barns achieved great lengths.

As early as 1710, Swiss Mennonites came to an area now known as Willow Street, three and a half miles south of Lancaster City. Amish settlers arrived by 1729 because of the extensive rich limestone soils. By the end of the eighteenth century, the county had become an agricultural powerhouse. Some sources contend that Lancaster County has the best soils in the world. Early settlers knew this and gravitated to Lancaster County for its relatively mild climate and remarkable soil.

High crop yields probably spawned the German two-level bank barn or forebay barn, which greatly expanded storage capacities of the considerably smaller one-level Ground barn that likely first occupied most of the hearth counties. The fact that so many barns over 80 feet in length existed in the county suggests that the demand for two-level barns and expansions occurred earlier in Lancaster County than in any of the other counties. The greatest concentration of the earliest forebay barns occurs in the Lampeter area just northeast of the earliest German settlement area in Willow Street.

## Piedmont Lowlands as the Most Productive Soil in the State

It seems that the earliest German settlers sought the most productive soil they could find. Lancaster County is located almost entirely within the Piedmont Physiographic Province. The area consists of three sections—the Piedmont Uplands that occupies the southern third of the county, the Piedmont Lowlands in the middle third of the county, and the Triassic-Jurassic Lowlands that occupies the upper one-quarter to one-third of the county. Easily eroded rocks such as limestone, dolostone, and phyllite underlay the lowlands. These rocks are relatively younger than the surrounding uplands and are likely the result of a quiet stretch of shallow sea deposition.

In *The Atlas of Pennsylvania* (1989), edited by D. J. Cuff and others, it is stated that little topographic relief and certain well-drained calcareous ultisols (strongly leached acid forest soils) make the Piedmont Lowlands of central Lancaster County some of the best agricultural land in the eastern United States.

In "Analysis of temporal variation and species-site relationships of witness-tree data in southeastern Pennsylvania" (2000), Bryan Black and Marc Abrams make the following comments: –

> "Early colonists purchased large parcels of the most desirable land in the Piedmont Lowlands and quickly established farms. By 1740, settlers began to experience a shortage of land in the Piedmont Lowlands as reflected by a sharp decrease in the number and size of (land) surveys and a dramatic increase in land prices. As survey dates indicate, settlement shifted to the less desirable Piedmont Uplands in the 1750s and then to the Triassic-Jurassic Lowlands in the 1760s. By 1770, land prices had quadrupled from what they were in 1736."

The area in and around the City of Lancaster occupies the area of the Piedmont Lowlands. In view of the above two studies, it is little wonder that Lancaster is one of the oldest inland cities in the United States, originally settled in 1718. The studies strongly support the idea that the soil (along with the general climate) was the predominant factor in attracting settlers.

## Seven Mennonite Families Come to Lancaster County in 1710

In his "1710 Pequea Settlement Tour Resource Information Booklet," Samuel Wenger cites seven Mennonite families who came from Europe to Philadelphia in 1710. Before arriving in the city, they were apparently aware of the magnificent productive soil that fostered the great forests of oak and a few other tree species in part of the area that later became Lancaster County—that is, the middle third of the county. A few travelers, among them William Penn, had made a trip to the area around 1700 and spread the word about the land's wealth. The few families procured a land warrant for 6,000 acres. This land was comprised of an area six miles by three miles from Willow Street to Strasburg. It was James Logan, the land agent for Penn, who said, "The Mennonites got the best land in our colony." The Pequea Creek and the Conestoga River provided the best soils. These areas gave rise to the earliest settled lands in the county. The 1719 Christian Herr house is just east of Willow Street.

It is interesting to note, as stated in Catherine Keever's study, "Distribution of Major Forest Species in Southeastern Pennsylvania," that white oak (*Quercus alba*) was located in the best farming area in Lancaster County. It was white oak (and a few other local oak species) that stood so prominently as the tree of choice with barn and house builders in the county from the first third of the eighteenth century onward. Witness this, for example, in the Christian Herr house dated 1719.

How prominent was white oak in the Piedmont Lowlands? One white oak stand that was cut before the start of the McKeever study was examined for twelve years. The trees were located on deep limestone that permitted the best growth of this tree type. Most of the abundant white oak trees were more than a meter (39.4 inches) in diameter. When the trees were cut, annual growth rings were counted and the white oaks were found to be about 250 years old. Such forest-grown trees are nearly unobtainable today in southeastern Pennsylvania. German settlers, of course, were no strangers to oak trees, as this species was common in the Palatinate in southwest Germany where many had come from.

At the present time, the exact time that Sweitzers first appeared on the landscape can't be ascertained. We can't entirely rely, for example, on a 1734 citation to "a bridgeway to the barn floor as reported in the 1915 History of the Church of the Brethren." This reference may or may not have been to a Sweitzer. The use of the word bank may not have been the same as we normally think of it today. Even one-level Ground barns may have been partially placed into banks. Witness the extant Kline stone and log Ground barn near Bowers in Berks County that is placed into a distinct slope or bank of ground.

The Isaac Long stone and frame Sweitzer in Manheim Township was long thought to be one of the earliest forebay barns in the state. The carved date of 1754 on a stable door frame is not a reliable indication that the barn of the type was built at that time. Ensminger and I have done extensive study of this barn for nearly twenty years and nothing definitive has been determined. It seems that the date is indicative of another building, perhaps a barn of some kind, at the early settled homestead, and that this building was greatly expanded later in the eighteenth century. Examination of the Long barn continues.

Alan Keyser, a German-American cultural scholar, told me several years ago that he once saw a log Sweitzer in the Manheim area north of Lancaster City a few decades ago. On the original metal strap at the front of the horse stable door was a carved date of 1765 or 1769 (he could not recall which). This date and the barn itself have never been verified. The oldest definitely dated Sweitzer reported by any barn observer in the last few decades is the Henry Beiler 1770 stone Sweitzer.

## Log Sweitzers

The earliest Sweitzers were probably made of double log cribs from timbers from old-growth forests. Some logs in surviving Sweitzers approach two feet thick and are seen at the bottoms of the cribs to add stability. The rare Herman double log crib barn (original section) with early style *liegender Stuhl* roof support and stone walled one-bay addition in East Lampeter in Lancaster County may be one of the earliest Sweitzers in the state. The log section may have been built between 1770 and 1790.

Log Sweitzers were built throughout the second half of the eighteenth century and the early nineteenth century. The latest Sweitzer built at least partially of logs is an 1856 structure in northern Lebanon County.

One of the most intriguing characteristics of log crib construction is its corner notching. Two basic types of notching characterize log barns in the hearth: inverted

An interior view of the one log crib in the 1796 log Sweitzer in Lebanon County shows the size of logs in this late-eighteenth-century barn.

The dovetail corner notching in this 1796 Schaefferstown Sweitzer is prevalent in the western half of the hearth area.

This Dauphin County log Sweitzer contains an excellent example of dovetail corner notching.

V-notching and dovetail notching (DT). A remarkable cultural phenomenon has occurred in regard to the distribution of these notches in log barns (and houses) in the study area. DT—either half dovetailing or full dovetailing—occurs in eighty percent of the barns in Lancaster, Lebanon, and Dauphin Counties, and the western half of Berks County. East of this area to the Delaware River, V-notching predominates.

Inverted V-notching in this log Ground barn with steep angles, also known as tear-drop notching, is often seen in pre-1800 log Sweitzers in the hearth.

This rather strict geographic demarcation of corner notching types seems to cut across time periods. That is, for example, the basic ratio of V-notching versus DT notching in the east area in the 1760 to 1800 era is roughly the same, for example, as in the 1800 to 1840 era.

What cultural forces held these corner notching methods in place? It may be a poignant example of the Wilbur Zelinsky principle of First Effective Settlement, which states that the pioneering culture in a particular area established the building style that subsequent generations would emulate.

The very large log and frame Sweitzer in northern Berks County has corner posting. Such posts in log barns are either mortised or slotted to receive the logs of the walls.

## Log and Frame Sweitzers

About a dozen combination log and frame Sweitzers have been identified. Their mows or cribs have either two or three log walls with the remaining wall(s) of frame. In every case, the wall adjacent to the wagon bay is frame construction. When only two log walls are present, frame walls are adjacent to the wagon bay and the front forebay. It seems likely that most of these barns were built post-1810. Frame construction adjacent to wagon bays is present in these combination barns likely because frame units or bents provide flat surfaces on which to attach the boards of mow-stead walls. Log walls do not provide even surfaces.

## Stone Sweitzers

Stone Sweitzers were built two or three decades after the first appearance of log Sweitzers—likely in the last third of the eighteenth century. As old-growth forests waned and farming became more lucrative, builders began using stone, which lasted much longer than log structures but was more labor-intensive. The earliest of these Sweitzers had stone-to-peak at the end walls and stone on the rear walls in line with the end bays or mows. In addition, it seems that a majority of these stone Sweitzers had stone front walls, that is, the walls at the rear of the front forebays. Most of these barns have distinctive wall openings such as early-style splayed loopholes or vertical slits (*balistratia*) or later type louvered windows. Such architectural elements are good general diagnostic dating tools. (See chapter on dating.) Their heyday was 1790 to 1820. For such an iconic example of Germanic vernacular architecture, this is a relatively short time frame.

Most of the later stone Sweitzers, built after 1810 or 1820, have stone on their end walls to the eaves level. This decreasing use of stone on the end walls marked

The Sweitzer in Upper Frederick in Montgomery County is a rare late-eighteenth-century frame barn dated 1795.

the gradual transition to all-frame forebay barns. During the mid-nineteenth century, wood boards were produced at sawmills at increasingly cheaper rates. Likely the timber sources were from second-growth forests. Plus, stone walled barns became increasingly cost prohibitive.

Likely, most of the stone Sweitzers in the greater Lehigh (occasionally seen) and Saucon Valleys (rare) have either vertical slits or louvered windows. Most of these barns were built prior to 1830. However, in the western half of Berks County and areas west to the Susquehanna River, a number of Sweitzers retain these wall openings well into the 1840s or even 1850s.

## Frame Sweitzers

The 1798 Direct Tax shows that frame Sweitzers were present in most of the hearth. In Upper Milford Township in Lehigh County is an extant rare stone and frame circa 1780 Sweitzer. Likely, the earliest dated frame Sweitzer that survives is of three-bay construction with a carved date of 1795 (with two additions) at the Sunrise Mill Historical Site in Upper Frederick, Montgomery County.

Still another early frame Sweitzer, circa 1790, later converted to a one-level barn (without its basement), is in New Hanover Township in Montgomery County.

Another frame Sweitzer is a two-bay construction in Spangsville in the Oley Valley in eastern Berks County. At least ten frame (in whole or in part) Sweitzers have been recorded in Lehigh County.

## Variant Log Sweitzers

The earliest Sweitzers, those of log construction, likely had no rear eave wall extensions. In certain post-1790 barns, these extensions, usally made of stone, are referred to as out-sheds and were a response to changing farm economies. Log Sweitzers with such out-sheds are known as variant log Sweitzers. With the addition of out-sheds, the end walls' classic asymmetry shifts to a more symmetrical roofline.

Variant log Sweitzers usually lie outside the hearth area, particularly in Somerset County; however, a few have been located in Lebanon County. The earliest examples are found in Lebanon County and nearby Cumberland County. The Monroe Valley barn was the finest one, probably built in the late eighteenth century, and was dismantled in early 1996. This structure apparently had several massive timbers up to 26 inches

The Funck variant log Sweitzer in Swatara Township, Lebanon County, has two non-original frame end bays and non-original frame rear bays or out-sheds.

The circa 1820 transitional double log crib Sweitzer in Pike Township, Berks County, is half chestnut, half oak, and retains its original rear wall wagon doors.

tall. The similar-type Olesh barn in western Berks County has one or two logs just shy of two feet thick. Curiously, many log barns (and log houses) have large timbers, usually of oak. These are found in the western third of Berks County and nearby Lebanon County. The Funck log Sweitzer in Swartara Township in Lebanon County has both end bays and two rear outsheds that are not original to the barn.

The rear extensions of Sweitzers may lend some fundamental insight into the fact that the early Sweitzer forebay barn did not meet the farmers' basic needs. When these variant barns were first introduced, other style bank barns had not yet appeared. The rear additions were probably used to store hay and straw or grain.

## Transitional Sweitzers

One rather rare type of the barn style, the transitional Sweitzer, has been identified on the landscape. This barn type is closed at each end wall, that is, the front edge of the basement-level stone wall aligns with the front forebay wall. Some of these barns are half open where the basement stone wall at one end is even with the stable wall. Twenty-two transitional Sweitzers have been identified and are constructed of stone, log, frame, stone and log, and stone and frame. These barns may have been a transition between the classic Sweitzer and the later Standard barn.

One variation features full stone on the end walls. The stone replaces the typical wood framing on the ends of the forebay appendages. An example is a two-section, eighteenth-century log Sweitzer in East Lampeter in Lancaster County. A 16-foot-long, one-bay addition features an end wall that is fully stone from the front edge (in line with the front forebay) to the roof.

The 1816 brick Sweitzer with rear out-sheds and rare flush frame ramp shed is the earliest dated barn of its type; located in Rapho Township, Lancaster County.

## Brick Sweitzers

A number of brick Sweitzers appear in the northern half of Lancaster County. They are common in areas outside the hearth, such as Franklin County. Some of these post-1820 structures are quite large. The earliest known dated brick Sweitzer, of 1816 vintage, is in Rapho Township in Lancaster County. These Sweitzers were likely built up to the Civil War and possibly later.

## Dating Sweitzers

Most of the recorded Sweitzers, more than 250 barns, are in the hearth area. With the single exception of Lehigh County, no systematic survey has ever been done to locate all the Sweitzers. I have recorded about 340 Sweitzers in twenty-four counties. Likely 500 other Sweitzers or more are still to be recorded in the state.

Lancaster County likely harbors at least 300 Sweitzers. Lebanon County probably has 150 to 200 Sweitzers. Berks County may have 150 or more. Farther east, fewer and fewer Sweitzers are encountered. To date, more than fifty Sweitzers have been identified in Berks County, but few have been seen in the eastern half. Lehigh County has thirty-four recorded Sweitzers and Northampton County has just two. Five Sweitzers are known in Bucks County, while three Sweitzers have been seen in Montgomery County.

In the past forty years, more than 200 barns of all kinds have been counted with their dates intact—carved

Opposite Page: The 1784 Knabbe stone-to-peak Sweitzer in Oley Valley, Berks County, is one of the oldest dated barns. This gem has one of the steepest roofs of any Sweitzer and has a German roof structure.

The Isaac Long Sweitzer in Manheim Township, Lancaster County, is a two-section structure and has mystery. The older section is dated 1754. The newer section as a full Sweitzer may have been built in the 1760s.

in stone, brick, or wood, and painted with lampblack—on timbers or boards. Fifteen of those are Sweitzers. Perhaps two to three dozen other barns not yet seen also date from the eighteenth century.

Of the dated Sweitzers, thirteen are stone, one is log, and one is frame. The earliest dated stone Sweitzer was built in 1770. This date, however, is questionable, as the date-stone is placed strangely on an end wall and accompanied by another distinctive stone that has no decipherable numbers or letters. One barn has a stable wall door frame with a carved date of 1772. Three other stone Sweitzers are dated in the 1780s. Seven Sweitzers date to the 1790s. One frame Sweitzer is dated 1795, and one log Sweitzer (dismantled in 2007) is dated 1796. The Isaac Long Sweitzer is stone to eaves level at both end walls. What seems to have been a separate structure with a date of 1754 on a lintel was greatly expanded, resulting in a 108-foot-long Sweitzer that now stands at the farm. The evolution of the Long Sweitzer is a complex affair, and the details have not been worked through. This has been an exciting project and a potentially rewarding one, as it may reveal some early history of Lancaster County Sweitzers.

## Nineteenth-Century Sweitzers and the Demise of the Sweitzer Barn

The basic type of the Sweitzer was well suited for the last quarter of the eighteenth century and into the first few decades of the following century. But then the basic configuration, with its prominent front appendage or forebay, became outmoded after about 1810. Expanding farm economies then demanded two things – greater storage capabilities and efficiency. The result was the proliferation of the second forebay barn type, the Standard barn. Many farmers had to adjust during the first few decades of the nineteenth century, or they would have been unable to compete in the marketplace.

Of course, some farmers chose to hang onto the Sweitzer type, because they had small farms or wished to continue a familiar tradition. Some of the latest Sweitzers, many of log construction and built in the 1840s, are in the Mahantongo Valley, which straddles Schuylkill, Northumberland, and Dauphin counties, including several adjoining valleys.

Sweitzers were still being built in the hearth area after 1840, particularly in the northern half of Lehigh County. A log and frame Sweitzer dated 1856 occupies the foot of Second Mountain in northern Lebanon County. Despite a rugged terrain that was difficult to till, the barn is large at more than 80 feet long. It seems unlikely that Sweitzers were constructed during or after the Civil War era.

## A Few Outstanding Sweitzers

### Stahler Stone and Frame Barn, circa 1780

While southern Lehigh County does not have a high concentration of Sweitzers, it has one of the most unusual examples of the type. The Stahler barn, on a backcountry road in Upper Milford Township, has a timber main section that looks shockingly like medieval construction. The front forebay, except for the cantilevered floor beams, was lost about one hundred years ago.

An 1856 log and frame Sweitzer in Union Township, Lebanon County, is the latest built dated log Sweitzer.

This rare stone and frame Sweitzer is the only definite eighteenth-century barn in Lehigh County. It has medieval-like timber framing and a German roof structure. The frame barn section (right) is a much later addition that eliminated the original forebay section.

Nonetheless, it is a local historic landmark and the only known definite eighteenth-century barn in the county.

The Stahler barn is a rare eighteenth-century frame-and-stone structure. When this condition is prevalent, it typically exists in either Sweitzer form or as a one-level Ground barn with stone on the lower half of the exterior wall and frame on the upper half. A number of combination barns are seen in the 1798 Federal Direct Tax. In other barns, the upper halves were log. We don't know why these combination barns were constructed. In the case of Ground barns, the upper halves, where crops were stored, likely allowed for greater ventilation. But the entire end bays of Sweitzers were used to store crops.

The Stahler barn has an original Sweitzer section and an early one-bay stone and frame addition. Only the Sweitzer section is discussed here. The Sweitzer end, which has a steep roof, is 56 feet, 8 inches long, and 30 feet, 2 inches wide. The front cantilever beams indicate that the forebay was about 7 feet wide.

The barn has three-bay construction, and one end bay is 7½ feet wider than the other. The barn has the early-style *liegender Stuhl* construction that consists of five roof trusses that include an accommodation of the frame end walls. (Stone end walls do not require the use of these specific trusses.) The three inner trusses were originally classic ones. Originally, the end wall trusses lacked end braces and lower straining beams. The timbers of the end trusses remain, while only one of the inner trusses has its original timbers. Its purlin plates are unusual, canted instead of square. Significantly, the plates extend horizontally about 3 inches beyond the original end wall. This aspect of plate extensions is seen in some eighteenth-century structures in the hearth. Several mid-eighteenth-century buildings at Ephrata Cloister in Lancaster County also have many plate extensions.

Finally, the barn has a unique complex of beams at the top of each eave wall. This includes English "sole-like" or horizontal beams about 18 inches long that join wall plates to inner parallel beams, adding an enigmatic element to the barn.

Upon entering this barn, most visitors notice the eave walls' prominent timbers. Such "over-engineered" or medieval-like timber assemblages have been seen in just a few other barns in the hearth.

## Bertolet-Coker Stone Barn, 1787

The Bertolet barn in Oley Township, Berks County, is a masterpiece replete with original architectural elements. Few barns exemplify the diverse aspects of Sweitzer traditions better than the Bertolet barn.

As an aside, a dated 1837 frame and stone Standard barn abuts the Bertolet Sweitzer. Built exactly a half century after the Sweitzer, it offers a tremendous opportunity to evaluate the progression of barn-building traditions over a fifty-year period.

The 1787 Bertolet Sweitzer in the Oley Valley is one of the finest of the type in the hearth. It has all manner of original features, including possibly the oldest *spriggel* bar and one of the oldest wagon floors.

The Bertolet Sweitzer has a unique construction feature called a *stickbalken*, or short transverse timber, at the bottom of each common rafter. Its exact function is unknown.

The Bertolet barn has one of the earliest carved dates on a mow-stead wall in the hearth—1818. Many other carvings are seen on the walls. A "Pennsylvania heart" appears on a mow-stead door leading to the granary.

The three-bay Sweitzer measures 65 feet long and 33 feet wide (without the forebay). The forebay is 6 feet wide. The front roof slope has a rare twenty-one-inch-wide remnant strip of early nineteenth-century shakes about 30 inches long with original beveled edges. The end walls and rear wall have splayed loopholes. The rear wagon doors (each half side) are of the rare *harr-hung* door variety. At the attachment side of the door half is a vertical pivot pole. The pin extensions at the ends of the poles are inserted into receptor spots. Two other rare exterior features include an original pent roof at the basement end wall and an original plank staircase at the front of the stable leading to the forebay.

The barn interior also features rare elements. The roof is an eighteenth-century *liegender Stuhl* and *stichbalken*, or short transverse beams, sit atop the side wall plates that meet the Common Rafters.

The granary in the forebay may well be the earliest in the state. In a near-miracle of survival, the ceiling, walls, entry door, and most grain bin areas are original. The forebay also has an original trapdoor with authentic wrought metal ware at the top of the staircase. Behind the trapdoor is a rare *hexel-Kammer*, or room for chopped up straw.

The Bertolet structure is one of the oldest barns in the state with a swing beam. About 7 feet above the wagon floor, it stretches the full width of the main barn section. A 5½-foot-wide alcove appears at the one side of the swing beam opposite the regular wagon bay floor that widened the floor space. Exiting the barn, unloaded hay wagons were able to make a wide turn, and tethered horses may have walked around the wide floor and performed threshing. The one mow-stead wall of the alcove has original beaded boards secured with wrought nails. A carved date of 1818 appears on one board.

The wagon floor has a number of wide planks nearly 2 feet across. Most barn planks don't exceed 18 inches.

The Bertolet Sweitzer is a barn that has largely refused to change for the past 225-plus years of its existence. We can hope that it may last for many generations, as it tells us so much about barn-building customs.

## Henneberger Barn, circa 1800

In the Welsh Run area of southern Franklin County, about eighty miles west of the Susquehanna River, the Henneberger stone barn is the Holy Grail of Sweitzers. It is ninety-five percent original, and although not in the hearth area, reflects the vast majority of features that existed in pre-1825 stone Sweitzers in the hearth. Hence its inclusion here.

The barn is a stone-to-the-peak four-bay structure and is 67 feet long by 34 feet wide with a 7-foot-wide forebay. It seems to retain its original rear wall wagon doors and stable wall single-height doors. Even the forebay has a rare original door at one end. Stone end walls and rear wall have vertical slits.

The barn has a Principal Rafter System. All structural roof elements are original. Astoundingly, about sixty percent of the original side-lapped German wood roof shingles survive. I have never seen this feature among the several thousand German-based barns I have visited.

The following aspects are nearly one hundred percent original: mow-stead wallboards, wagon floor planks, and the granary in the forebay including its entry door, flooring, grain walls, and ceiling and rear wall boards. This may be the most original 200-plus-year-old granary in the state.

Almost all of the basement's original staked mangers and troughs are intact. The one feeding alley has a rare two-section end-to-end wood box 18½ feet long with finely crafted wrought hinges. The original rear wall staircase to the second floor survives along with a rare remnant plank floor in one of the feeding alleys.

The McCoy family built the barn, and this barn is truly the real McCoy. Strangely enough, another nearly original barn, this time in the hearth, revealed itself in 2012 in Jackson Township, Lebanon County. The brick Standard barn dated 1859 was built by the Groh family some fifty years after the Henneberger barn appeared.

The Henneberger stone Sweitzer in Franklin County outside the hearth is the Holy Grail of 200-year-old German structures. It retains ninety percent or more of its original features on both floors.

Both the Sweitzer and the frame attached section at the Henneberger homestead retain most of their original German side-lapped shingles. Such surviving shingles are extremely rare.

## Isaac Long Barn, 1754 (two sections)

The Isaac Long barn is the most interesting and complex North American structure of its type that I have seen. It is in Lancaster County about four miles northeast of Lancaster City. I have visited this astounding barn more than a dozen times starting in 1992, and I am impressed anew every time I see it. I wonder if its intricacies will ever be solved. Where do you begin? For one thing, its steep pitched roof is one of the most classic roof silhouettes in the hearth area.

The barn seems to have been built in two sections. The first and much smaller section was erected in 1754 and later expanded to 108 feet, probably in the last third of the eighteenth century. The original section's stone end wall was heightened to the eave level with stone. Above that, frame was added to the peak. Many recycled timbers appear in both the original section and the added section. Much of the stable wall was altered, although three original doors survive in the 1754 section.

On the upper floor you can see the Germanic roof type with five early-style trusses, both classic and variant. The inclusion of early-style wrought nails in mow-stead wall-boards, however, are a strong indication that the later addition was built before 1800.

To further complicate our understanding of the barn, the addition's ceiling joists consist of beams recycled from a log structure that was likely on the homestead. Other recycled timbers are also found throughout much of the barn.

This barn has enough conflicting components to confound anyone, but it's a good challenge for everyone including traditionalists. The only real way out of the quandary is to dendro-date many dozens of timbers in both barn sections.

## Christian Meyer Barn, 1772

The Christian Meyer barn stands far back on an old farm lane in rural West Earl Township, Lancaster

County, but it is on the front line of the earliest remaining Sweitzers in the state. The date 1772 is carved on an original stable door frame along with the initials C. M. for Christian Meyer, who settled here in the second half of the eighteenth century.

The Meyer barn has stone to the eaves level construction and the one end wall has splayed loop-holes. The far end addition is frame and replaced the one original far end wall of stone. With four bays, the barn is quite long. The basement has been dramatically altered in the past seventy-five years, and only four or five original full-barn-width ceiling joists remain. They are a clue that the barn is a Sweitzer, as the original forebay is gone, along with most of the stable wall.

A banked structure, the barn now has a Common Rafter System that consists of hewn rafters. The purlin plates and queen posts are canted, and transverse bents have a few recycled timbers.

Despite the presence of all the loft timbers, it seems likely that a great reconfiguration of the original timbering occurred. Several aspects indicate this. A pre-Revolutionary War barn would have a *liegender Dachstuhl* roof. The bottom ends of the common rafters have post-1810 "shaped" (curved) tails, whereas pre-1810 rafters are usually straight. Also, canted queen posts would not normally be in an early barn. It would most likely have had a *stehender Dachstuhl* (thick vertical queen posts) or the German *liegender Dachstuhl*. In addition, the bents would not have dropped upper tie beams. Instead the bents would have such ties at their tops. These later-style construction elements indicate that the upper level was greatly changed.

Two of the bents have recycled timbers that function as long braces. Each beam has two long, thin, empty mortises. Each beam closely resembles a lower tie or straining beam common in *liegender Dachstuhl*

The Christian Meyer stone and frame Sweitzer in West Earl in Lancaster County is an apparent remnant structure originally built in 1772. Construction on the upper floor and much of the basement was altered. The original forebay appendage was replaced by a late date front shed (at left).

construction. It would seem that the two timbers originated from the original roof structure.

The barn presents a puzzle of construction elements. Nevertheless, the barn was unquestionably built early on, reinforcing the consensus that some Sweitzers were constructed before the Revolutionary War. It would be fortunate if a few more pre-1775 barns would show their faces in the hearth.

## Summary

The hearth area where Sweitzers abound to this day offers exciting opportunities to experience the earliest remaining examples of forebay barns. More than 500 examples survive, offering us the ability to study their evolution and variations. A world of discovery awaits future scholars who take time to seek out all the remaining Sweitzers and add to one of the greatest stories of barn-building traditions in the East.

# Chapter Six.
# Standard Barns

The magnificent 1849 dated Wagner barn in Lower Saucon Township in Northampton County is one of the wonders of the Standard barn world. The craftsmanship exceeds all other barns in the hearth area, bar none.

While Sweitzers may have an earlier and more romantic history, the Standard barn reigns supreme as the most successful forebay barn style, number-wise, built in Pennsylvania. Tens of thousands of Standard barns were built between 1800 and 1925, and many have retained their usefulness. Farmers still use Standard barns to store crops on the upper floor and stable farm stock in the basement. It seems a near certainty that Standard barns will persist several hundred years from now.

No historian knows where or how this almost ubiquitous barn style originated. Since Standard barns have no European antecedents, scholars assume they evolved from the Sweitzer. However, this is uncertain, and we are still left with the startling realization that a

The dated 1819 stone-to-peak Standard barn in the Riegelsville area of Bucks County has the distinctive regional circular loopholes.

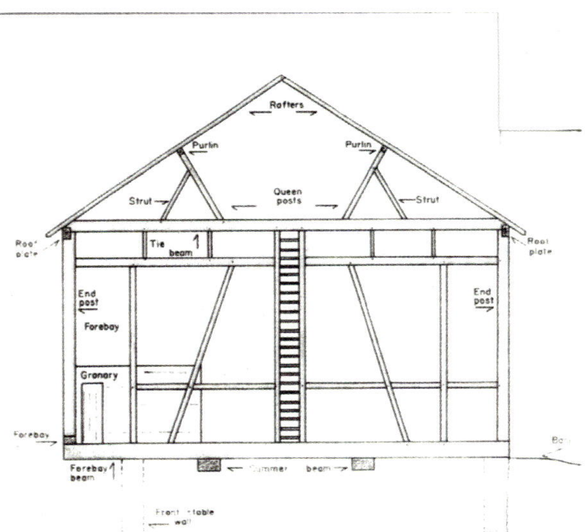


Framing of a transverse bent in a Standard barn includes a roof structure with canted queen posts, and framing below with upper tie, upper and lower girts, posts and studs, long sill to girt braces, and built-in ladder. At barn front (left side) is the forebay. Below the framing is a basement. The roof silhouette is symmetrical. (Courtesy of Ken Easterday)

comprehensive answer lies in the future. In the meantime, much has been learned about this common forebay barn in the past few decades.

We will examine their exteriors from bottom to top and then the interiors in reverse order. Foundation walls were nearly always stone. Upper floors were built of a few materials including stone, frame, a combination of stone and frame (on end walls), and occasionally brick. Rarely were they made of logs.

## Exterior

A Standard barn is a two-level banked forebay structure. Most of the barns have gable ends, but some post-1880 examples have gambrel roofs. Several two-level bank barn styles exist in North America, but the defining feature of a Standard barn is a forebay. In this critical aspect, Standard barns are similar to Sweitzers, except that Standard barns have symmetrical rooflines—each roof slope is the same length. Standard barns typically have narrower forebays than Sweitzers. Eighty percent of Standard barns have forebays 4 to 5 feet wide.

Exterior dimensions vary widely by location and number of bays, though some three-bay barns are actually longer than some four-bay barns. The following data come from two surveys of seventy-four barns. Seven barns have two bays, fifty-one barns have three bays, fifteen barns have four bays, and one barn has five bays.

The first survey includes a random selection of forty-two Standard barns documented since 2002 in all counties of the hearth except Philadelphia, Delaware, and Dauphin. The numbers are likely close to what would be generated from a controlled study.

Berks County: Ten barns averaging 57 feet by 34½ feet. Four barns have three bays and average 43½ feet by 31½ feet. Six barns have four bays and average 66 feet by 37 feet.

Bucks County: Three barns have three bays and average 51 feet by 33½ feet.

Chester County: Three barns averaging 37 feet by 28½ feet. Two of the barns have two bays averaging 31 feet by 25 feet, and one barn has three bays and measures 50 feet by 36 feet.

Lancaster County: One barn with four bays measuring 56 feet by 36 feet.

Lebanon County: One barn with three bays measuring 55 feet by 38 feet.

Lehigh County: Thirteen barns averaging 53 feet by 34 feet. Nine of the barns have three bays and average 50 feet by 33 feet. Three barns have four bays and average 69 feet by 36 feet. One two-bay barn measures 35 feet by 34½ feet.

Montgomery County: Two barns with two bays averaging 38 feet by 30 feet.

Northampton County: Nine barns averaging 39 feet by 31 feet. Two barns have two bays and average 28 feet by 27½ feet. Seven barns have three bays and average 42 feet by 32 feet.

The average two-bay barn (five barns) is 30.6 feet by 27.8 feet. The longest two-bay barn is 35 feet long and the widest is 34½ feet wide. The shortest two-bay barn is 28 feet long and the narrowest barn is 24 feet wide.

The average three-bay barn (twenty-seven barns) is 46.4 feet long by 32.7 feet. The longest three-bay barn is 65 feet, and the widest barn is 38 feet (two barns). The shortest three-bay barn is 28 feet long and the narrowest three-bay barn is 24 feet wide.

The average four-bay barn (ten barns) is 60 feet long by 36.4 feet wide. The longest four-bay barn is 80 feet long, and the widest barn is 41 feet wide. The shortest four-bay barn is 56 feet long, and narrowest four-bay barn is 28 feet wide.

The second source involved thirty-two barns in the greater Hellertown area of Northampton County in the period 2007 to 2014. Twenty-one barns were located in Lower Saucon Township in Northampton while six barns were in either nearby Bucks County or Lehigh County.

Here are the dimensions of Saucon Valley barns with two to five bays:

- Two-bay barns: Nine barns averaging 30 feet by 27 feet.
- Three-bay barns: Twenty-four barns averaging 48.7 feet by 34.8 feet.
- Four-bay barns: Five barns averaging 71.8 feet by 40½ feet.
- Five-bay barns: One barn measuring 85 feet by 40½ feet.

In that cross-section of seventy-four barns three- and four-bay barns are the most abundant. It seems likely that this would be consistent throughout the hearth area. Certainly a number of five-, six-, and even seven-bay Standard barns have been identified, and their lengths often exceed 90 feet, and even, remarkably, 100 feet. The absolute upper limit of a Standard structure in the hearth is the 150-foot-long by 55-foot wide Kline barn circa 1890 in Jackson Township, Lebanon County.

In the two studies, the exterior dimensions of two-bay and three-bay barns closely match, but their length

The stone-to-eaves-level Standard barn in Greenwich Township, Berks County, retains all its original forebay wall vertical siding along with all its original forebay wall doors.

varies by 12 feet and their width by almost 4 feet. These differences are fairly common and have no particular significance.

## Front Forebay Walls and Stable Walls

The fronts of Standard barns consists of two surfaces: the upper front wall of the forebay, and the recessed stable wall beneath it. The upper wall is always framed and is covered with vertical boards; some barns retain their original boards. The boards typically run in single pieces vertically, 16 to 18 feet. Boards are 10 to 20 inches wide. Many of the boards probably originated from old-growth forests. The earlier (pre-1850) barns likely more often had front walls of pine; after that they were hemlock. In the middle area of the walls, in line with the wagon bays, are winnowing doors that are also occasionally original. These doors were supposedly left open for fresh air during threshing or winnowing. In the winter, these doors were likely in a closed position. In addition, other doors are sometimes seen on front walls in line with the end bays or mows of the barns. These doors may be a part of the front walls of granaries that were often placed in the forebay area.

Due to its very tall front forebay wall, the Douglass Township Standard barn in Montgomery County has a rare double row of barn stars. The barn also has a rare area brick stable wall.

Occasionally certain barns have their front walls adorned with barn stars or hex signs of various types. It is known with near certainty that barn stars were first applied to barns in the hearth area in the first decade of the nineteenth century, perhaps earlier. But the predominant barns of this era were Sweitzers and not the Standard barn style. Many people regard barn stars as interesting decorative touches to barns.

*Peiler eck*e (pier corners) are often seen at the front corners on the basement level. These are stone ell-formations that helped to reinforce the stone end walls. Such formations are seen from the ground level to the top of the eave walls in stone barns. In frame barns they only appear on the basement level. Some barns have a *peiler eck* at one corner of the barn and a straight wall (with no ell) at the opposite corner. On the house-facing side at some homesteads, the barn's *peiler eck*e often have associated cupboards with single-height entry doors, which temporarily stored harnesses on stout pegs in beams set into the stones walls. *Peiler eck*e at the opposite ends of these barns were open areas that had only occasionally had cupboard sections (with doors).

The Morrison stone Standard barn in East Brandywine Township in Chester County has a prominent *peiler eck* at each of its front corners. The inner side has a rare curved aspect.

A small percentage of barns have harness peg beams toward the top of the stable walls. Some barns retain their original pegs. In a number of Standard barns, no *peiler eck*e are seen at the front corners. Instead, the stone end walls of the basement level extend straight out and no ell formations are created. In other barns, stone walls angle out for greater support, but actual ell-formations are not present. Barns with stone ells and those without ells (straight wall extensions) are called closed Standard barns. In the Ensminger system, they are Class 2, Type A. When such stone walls do not extend beyond the stable walls on basement levels, the barns are said to be open Standard barns, or Class 2, Type B. Some Standard barns are open at the one end (usually at the house end), and closed at the other end.

## Stable Walls or *Stahlwand*

Stable walls or *stahlwand*, depending on dimensions of the barns, have a variable number of doors. These doors are usually split into upper and lower halves. Often, the upper and lower door halves are even in height. When they are not, the upper door halves are several inches shorter than the lower halves. Doors vary in their widths, from about 36 to 42 inches or sometimes considerably more in barns of the last quarter of the nineteenth century. Often, the upper halves, especially in warm weather, were left open for ventilation. A fair number of Standard barns retain their original stable wall doors. The doors have a number of construction details worthy of a small treatise, but too numerous to mention here.

The stable wall doors do not always have uniform widths in any given barn. Doors for human entry were a few inches narrower than doors for animals. Human doors led into feeding alleys or *fudergong*. The door entries often had stone thresholds.

The original frames (planks) or jambs of stable wall door entries are often intact. They vary in width by a factor two to one—from 10 to as much as 20 inches. Many are 14 to 18 inches wide. Probably a majority of the time, the frames are tapered, wider at the outside. The reason builders tapered these frames is unknown, but theories have been proposed. The most common

The circa 1860 Ruch frame Standard barn in Lower Saucon Township in Northampton County retains its full complement of original stable wall doors.

The original stable wall of a Standard barn in Lower Saucon in Northampton County was removed when the front straw shed was added. The stable wall was repositioned in line with the barn's front corners.

is that the tapering saved wood—one could obtain more planks per volume of wood than if the planks had uniform thicknesses. It seems more likely, however, that the planks were tapered for aesthetic reasons. The tapering also created something of a "funnel effect" for farm stock to exit the basement through the doors. The frames are pegged at their upper corners. In many post-1860 barns, trim pieces were nailed to the top of the frames—often of oak but sometimes chestnut. Some door frames consisted of beams only a few inches wide.

Stable walls also had two to four barred windows, depending on the barn's size, for ventilation and light. The windows were interspersed among the doors. The tops of the windows were about in line with the tops of the doors. Each window had roughly eight horizontal bars. The bars are virtually always set on the diamond at 45 degrees. This angling is more aesthetically pleasing than bars set flat. The windows also have pegged upper corners. A fair number of the windows originally had shutters attached at their fronts. They were likely closed in the cold weather to retain heat. A few barns have vertical bars.

Many Standard barns have wagon or cart sheds at one end of the basement, as do some Sweitzers. They are either integral to the barn or have been added on several decades after the original barn was constructed. The sheds are fully open areas from the barn front to the rear wall and are 12 to 20 feet wide. Their exterior sides are stone or frame with short stone foundation walls. Some of these cart sheds have open rear areas where the carts can exit out the rear. Nearly always, the ceilings consist of longitudinal (end to end) joists or beams. Often, windlasses are present on the ceilings. Long farm ladders were probably suspended there. It is difficult to know just when cart sheds were first incorporated into these barns.

## Original Versus Non-Original Stable Walls

All Standard barns originally had stable walls recessed behind the front corners. But some Standard barns have not retained these recessed walls. Instead, they have stone or frame walls that align with the forebay wall. Such walls were almost always built to accommodate

The stone Standard barn in the Pektor Hill area of Williams Township in Northampton County has the distinctive regional circular loopholes built within about ten miles or so of the Delaware River.

different uses. Often the original recessed stone walls were dismantled and brought forward several feet to create more space. This condition can be confusing to new barn observers and even experienced ones. A bank barn in Lower Saucon Township in Northampton County provides a good example. The barn at first was thought to be an English Lake District barn due to the fact that the stone stable wall was in line with the barn front corners. Other aspects seemed to confirm this initial assessment. But a closer look clearly showed that the barn was indeed originally built as a Standard barn.

A number of Up-country posted barns have stable walls in the main rear section (Standard barn section) that were moved to the barn front when the front straw sheds were added. Sometimes, original stable wall doors were retained in the wall relocation procedure, but occasionally new doors were installed in the new walls.

## End Walls

Most Standard barns have stone foundation end walls 7 to 8 feet high. On the gable end wall facing the house, one or two doors appear within a few feet of each other toward the rear wall corner, along with one or two small barred windows. Doors are rarely seen on the opposite or far end wall. Two or three barred windows are normally included on these far end walls.

In their evolution, end walls above basement levels in Standard barns followed this sequence: stone to the roof peak, stone to the eaves level, and then frame. Stone-to-peak barns often have wall openings in the form of vertical slits (splayed loopholes) arranged in three to five rows, or louvered windows in double or triple rows. Occasionally, other wall openings may appear. Barns with stone to the eaves level have these openings but to a far less extent. A number of barns in the Saucon Valley demonstrate a distinctive regionalism where end walls have circular loopholes. The holes, 9

The frame Standard barn in North Whitehall Township in Lehigh County has stone-filled put-log holes on the basement level and the stable wall. Scaffolding poles were inserted in these holes during construction.

to 10 inches in diameter, are outlined with nine or ten bricks that radiate out from the hole. Stone barns built in the Lehigh and Saucon Valleys after about 1850 seldom have any wall openings. This is not the case in barns in the west section of the hearth. Thus, the openings serve as good general diagnostic dating tools.

## Put-Log Holes

One of the most interesting aspects of stone walls in Pennsylvania barns is the placement of so-called put-log holes. These small areas, about 4 by 4 inches, were placed strategically in stone walls to hold scaffolding poles as the walls were being built. The lowest holes, all in a row, were about four feet above the ground. A few frame Standard barns even have these holes on their stone foundation walls. Higher rows were placed at 4-foot vertical intervals. After the scaffolding poles were removed, the holes were often later filled with stone.

## Rear Walls

The rear walls of stone-to-peak Standard barns have stone from ground level to the tops of the walls at either side of the centered wagon doors. The end wall opening types mentioned above also appear on the rear walls of barns built before 1840 or 1850.

Rear wall wagon doors always swung outward, at least those built prior to about 1875. A few types of wagon doors were constructed, chief among them the common batten type and the earlier frame type. (Details of wagon doors are discussed in another chapter.) Pre-1830 barns rarely retain their original wagon doors. A fair number of post-1860 barns do retain their original doors. Some barns built after 1875 have their original sliding wagon doors.

Some large Standard barns have basement doors in the stone foundation rear walls, aligned with end bays in the upper level. This placement probably allowed farmers to access the basements when they were performing chores behind the barns.

Original wagon doors and the one human door at the one side are retained on this stone Standard barn near Pike Township in Berks County.

## Approach Areas to Rear of Barns

Approach areas to rear wagon doors vary in two principal ways. When barns were built into sides of hills, the wagon doors were at ground level. Thus, horse-drawn hay wagons could easily enter the barn. When barns were built on level ground, the excavated earth was used to create a ramp. These ramps typically had stone retaining walls and provided wagon access at the rear of the barn. Some of these stone-walled ramps are spectacular, such as the one at Camp Olympia off Cedar Crest Boulevard in Lower Macungie Township, Lehigh County. Another unforgettable ramp is at the Breckbile farm in Paradise, Lancaster County, where one ramp is nearly 100 feet long.

## Roof Coverings

Before roughly 1875, there were just two types of roof coverings: thatch and wood shakes. Thatched roofs were the earliest ones and were used into the early 1900s. It is difficult to know when wood shakes were first used, but they were well established in the eighteenth century. At first, German side lapped shingles (shakes) were used, and remnants (under later roof coverings) are rare. The remarkable Henneberger Sweitzer in Franklin County retains many of its original shingles under a much later covering. These early shingles were quite long, 30 to 32 inches or more, and made of red oak. Later-style shakes were the simple top-to-bottom overlapping type and were quite shorter, 20 inches or less.

It seems likely that German side-lapped shingles were still in vogue in the first quarter of the nineteenth century when Standard barns started to proliferate. But at some point in the 1820s or 1830s, these shingles were phased out in favor of riven wood shakes connected in a top-to-bottom fashion. Such second- or third-generation shakes still exist. One outstanding example is the Bertolet Sweitzer in Oley Township in Berks County.

## Interior—Upper Floor

The interior of all two-level banked barns, no matter their type, style, or geographic location, consists of an upper level or loft and a basement level. The following discussion focuses on the structural and architectural elements of each level. (See the chapter on Construction Elements for further details.)

### Roof Supports

Standard barn roof support structures include the three major types seen in Sweitzers. The most common is the Common Rafter System, a series of same-size rafters arranged in pairs. In any given barn, rafters are usually pegged at their top ends and over-sail the wall plates at their bottom ends. Most pre-1860 Standard barns have hewn (hand-formed) rafters; barns built after 1860 usually have milled rafters.

Another major rafter type is the Principal Rafter System. At least 120 Standard barns have been identified with this English roof system, and it is likely that three to four times more Standard barns have these roofs.

The prominent earthen approach ramp with a tall side stone wall to the Camp Olympic Standard barn in Lower Macungie in Lehigh County illustrates the manner in which loaded hay wagons entered many bank barns in the hearth area.

The Henneberger stone Sweitzer in Franklin County (outside the hearth) retains many of its original German side-lapped wood shingles. Likely many Standard barns used these long-style shakes from the early 1800s until about 1830.

The Morrison barn in Chester County retains its original hewn timbers in its Common Rafter System. The purlin plates are supported by canted queen posts.

The Schaefferstown area log Sweitzer had an excellent Principal Rafter System. Most Standard barns with this roof structure very closely resembled the Schaefferstown area roof system.

The Stapleton variant Ground barn with its completely original classic *liegender Dachstuhl* roof structure is very similar to the same style roof structures seen in fifteen Standard barns.

The Kerschner frame circa 1860 Standard barn in Heidelberg Township in Lehigh County has typical transverse bents complete with upper tie beam, lower level girts, braces, and built-in ladder.

A third roof type is the early German style or *liegender Dachstuhl* roof support, which is similar to the Principal Rafter type. About fifteen Standard barns have been discovered with this unusual roof type. Normally, this early roof style is not associated with nineteenth-century barns and certainly not with houses of this era. This roof system was far more labor-intensive than the Principal Rafter System. This is the reason the German roof system became outmoded after about 1800. The Industrial Revolution was fast approaching and efficiency was important.

The earliest Standard barn with the German roof support was the 1804 DeLong barn near Bowers in Berks County. This massive barn, more than 105 feet long, was destroyed by a tornado in late May 1998. A few builders included this roof type as late as the 1840s. This roof support includes true trusses where each truss has five timbers—two Principal Rafter pairs, a connecting cross tie or straining beam, and end braces. This is referred to as a classic truss type. Another type or variant truss lacks either end braces or a straining beam. Usually barns of the later era have variant type trusses.

## Purlin Plates and Queen Posts—Common Rafter Support System

Standard barns with Common Rafter Systems were often in need of purlin plates and queen posts. The English and German roof types did not typically make use of these two timber types.

## Transverse Framing Units—Bents

Transverse bents in Sweitzers were used often; they are even seen in a number of log Sweitzers. But the practice was far more pronounced in Standard barns, and more evolved. Recall that bents stretch the full width of almost all Standard barns, while in Sweitzers, bents never extended into the forebays. These bents were only present in the main or rear sections of non-log-built Sweitzers.

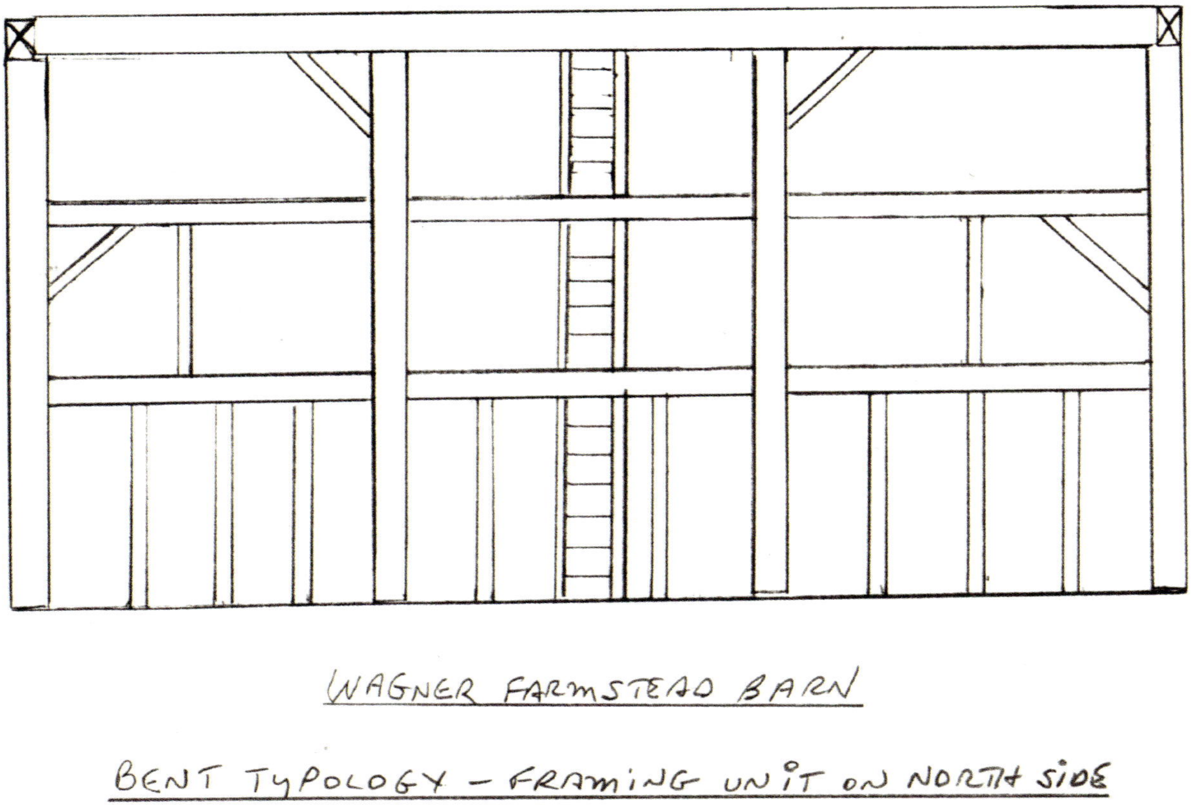


The Wagner stone-to-peak Standard barn dated 1849 in Lower Saucon Township, Northampton County, has standard transverse bent morphology. The bents consist of upper ties, upper and lower partial-barn-width girts, braces, and built-in ladders.

## Alternative Construction—Arched Longitudinal Beams

As of 2015, about ten Standard barns in the central third of Bucks County have been identified with this rare type of construction. Unfortunately, one of these barns located in Hilltown Township, was dismantled as early as 1976.

All Standard barns with these arched beams have three bays. The arched timber, set longitudinally, is about 8 feet above the wagon floor and stretches the entire width of the floor between a post of each transverse bent that flanks the wagon bay. This rather substantial beam is positioned about 12 feet to the rear of the front forebay wall. The arch or curve of the beam is seen on both its upper and lower surfaces, creating a pronounced visual effect. In four or five of these barns, the ends or tenons of the beams extend fully through the posts and terminate as salient tongues. The extended tenons, generally about 8 inches long, are double wedged, similar to a Dutch-American barn anchor beam. The beams vary at their midpoints from 12 to 14 inches in height. Most of the arched beams barns are oak.

A stone Standard barn in Doylestown in Bucks County has an unusual arched anchor-beam-like timber on its upper floor near the front forebay wall. A few barns in mid-county have such timbers.

At the front vertical surface of these beams that face the front wall, a series of equally spaced mortises house milled joists about 8 by 3 inches. The joists extend to a similarly arched horizontal beam (without extended tenons) that abuts the front wall. The two arched beams and joists in concert seem to have functioned for extra hay storage or mow area in the middle bay of these three-bay barns. Boards were placed on the joists as a place to store crops.

## Two Means of Shouldering

In almost all Standard barns of the pre-1910 or perhaps pre-1920 era, joinery in transverse bents involved the use of mortises and tenons. Timber ends have thin tongues of wood called tenons. They were placed into mortises (voided wood areas) of the timbers they were joined to. The manner in which beams, usually ties, were housed or "brought or let into" the mortised posts was called shouldering. Long pegs or wooden pins, also called tree-nails or trunnels, secure the joints. Such joinery, as well as shouldering, is seen in thousands of Standard barns.

## Built-in Ladders

Bents in Standard barns almost always have built-in ladders. Their fine details were executed in too many ways to mention here.

## Mow-stead Walls

Mow-stead walls are adjacent to wagon bays, whether the barns are two-bay, three-bay, four bays or more. Not all of these walls have survived in Standard barns.

One unique situation prevails in a circa 1800 stone-to-peak Sweitzer barn about two miles west of Leesport in Centre Township, Berks County. Amazingly, two mow-stead walls consist of 2-inch-thick horizontal oak planks. A few of the planks are nearly 18 inches wide and have tongues at their ends that are received into vertical grooves in the studs of the bents. The planks have marriage marks. How often such planked mow-

An altered Sweitzer in Centre Township in Berks County retains part of one of its rare original plank mow-stead walls. Oak planks are 2 inches thick. Some Standard barns likely had such mow-stead walls.

A nearly all original granary survives in a gable ramp entry Standard barn in Longswamp Township. This granary has a central passageway seen in several Lehigh Valley barns.

A small frame Standard barn in Lower Saucon Township in Northampton County retains most of its original *oberdenn* boards. These boards are very wide at more than 2 feet.

A frame Standard barn in Ruscombmanor Township in Berks County has most of its original *oberdenn* longitudinal timbers. Farm crops were stored above these beams.

stead walls were utilized in barns is unknown. Only two other barns, both of them English Lake District structures pre-dating1800, have such planked walls. It seems likely that some Standard barns had these mow-stead walls.

A nearly fully original hay hole structure survives in an 1850s Standard barn in southern Lehigh County. Hay was dropped through an opening to the basement so animals could feed.

### Granaries

Granaries were usually located at the front sides of the barns on the upper floor level. They occupied, in an end bay, all the forebay section plus a few feet to the rear of the forebay. Some have central aisles with grain bins at either side.

### Crop Storage Areas

Crop storage sections in Standard barns (and in Sweitzers) were basically confined to two areas on the upper floor level. The first were spaces in end bays and also areas 11 or 12 feet above the wagon floors known as *oberdenn*s. The end bays in any given barn are quite often of different widths.

In normal eave-wall-entry Standard barns, the long axes of the *oberdenn*s were fixed transversely, or eave wall to eave wall. In gable-entry Standard barns, however, the *oberdenn* storage areas were rotated ninety degrees, or end wall to end wall, creating a dynamic visual effect.

### Hay Holes or Funnels

Farmers needed a way to deposit crops from the upper reaches of mow areas to the basement. Thus they included hay holes that functioned like funnels. They were typically placed against the rear walls in an end bay or mow.

### Wagon Floor Planks

Wagon floors had to be constructed of heavy wooden members called planks that could support the great weight of loaded hay wagons. Many barns retain most of their floor planks in the wagon bays. Planks are normally 2 inches thick and up to 25 inches wide. Most of these planks are 10 to 14 inches wide. Many wagon floors have thin pieces of wood or splines set in grooves between the planks.

### Basements

Basements have two main areas of interest—the complex of ceiling beams, both joists and summer beams, and stables. The wall surfaces are often white-washed. These basement elements are fairly consistent from barn to barn.

### Orientation of Summer Beams and Joists in Gable-Ramp Barns

One of the special forms of the Standard barn is the Gable-Ramp Standard barn, Type F-1 in the Ensminger system. Here, the main wagon doors appear on the upper floor level of the gable wall. Barns of this type maintain the normal eave wall forebays. Hay wagons traversed the gable ramp at a right angle compared to the eave wall wagon entry. This created an interesting reordering of certain wood timber features. It required the rotation of three common construction features ninety degrees from their normal orientations. These involved the summer beams, the basement ceiling joists, and the upper level wagon floor planks.

Summer beams in regular eave wall entry barns are longitudinally oriented, end-wall-to-end-wall timbers in basements; they support the transverse joists above

A gable ramp barn in Longswamp in Berks County has ceiling joists below the wagon bay oriented longitudinally. This was for the proper nailing surfaces for the planks of the wagon floor.

them that span the rear wall to stable wall. Barns usually have one or two summers. Joists may number twelve to twenty-four or more. The joists support the upper level floors. The wagon floor planks are oriented longitudinally in these barns.

When loaded hay wagons entered barns, the floor planks on the upper floor were perpendicular to the path of travel, creating good traction. In gable entry barns, transversely oriented planks meant that the joists stretch side to side or longitudinally in the basement to create a proper nailing or pegging surface for the planks. Basement joists regularly run transversely, or front to back, in barn basements. In turn, the side-to-side orientation of the joists requires the summer beam(s) to be positioned transversely rather than in the usual end-to-end or longitudinal orientation. With the three

The DuPortail stone-to-peak Standard barn is the only identified barn of the type from the eighteenth century. It has a date of 1792.

An original-looking 1801 date-stone appears on a stone Standard barn east of Bethlehem in Northampton County. The interior timber framing, however, suggests a construction date in the 1830s or 1840s.

special orientations in these Gable-Ramp barns, a different re-ordering of planks and beams was seen.

Builders knew how to position wood construction elements for best use of farm machinery and vehicles. This is yet another case of form following function.

## Era of Construction

Standard barns were built between the 1790s and 1920s—a period of 130 years. In January 1995, I discovered a dated 1792 stone-to-peak Standard barn at the Duportail homestead near Devon in Chester County.

The presence of this barn was a revelation, especially because it was in an English-settled area. Standard barns are, at this point in our understanding of them, based on German barn-building traditions. The Duportail barn is the only eighteenth-century Standard barn ever seen. I also found a barn in Bethlehem Township in Northampton County that has a date stone of 1801. Its interior, however, suggests a barn built about forty years after that date.

Although many Standard barns have been lost in the last few decades, we have a general sense of when these barns started to appear. Eight Standard barns built in the 1800 to 1809 era have been identified in Bucks, Berks, and Chester counties. Several other dated and undated barns likely exist from this era.

Twelve dated barns have been documented in the next ten-year period, 1810 to 1819, in Bucks, Montgomery, Lehigh, and Berks counties. In addition, eleven dated barns were found from the 1820—1829 period, and likely there are more.

The number of barns documented in each of the three decades signify that the Standard barn style was a definite part of various local cultural landscapes in the first third of the nineteenth century in the hearth area, and that economic changes were already occurring before the Industrial Revolution started in earnest in the third decade of the century.

After about 1830, thousands of Standard barns were built in each decade until the late nineteenth century, and many are still functional today.

The following is a breakdown of Standard barns by their three major material categories: stone-to-peak, stone-to-eaves-level, and frame constructions:

- Stone-to-peak barns were the earliest structures and were occasionally built into the late 1850s or even early 1860s. I have seen one 1880 barn in Oley Township, Berks County and another barn dated 1884 in East Nantmeal, Chester County. Their heyday was from 1810 to 1840.

- Stone-to-eaves-level barns appeared around 1830 and were probably built into the 1870s and 1880s.

- Frame barns were generally built after about 1840 and appeared continuously until the second decade of the twentieth century. A few exceptions may exist of these barns built prior to 1840 or even 1830. However, according to the 1798 Federal Direct Tax, all types of frame barns were built in the eighteenth century. These were both Ground Barns and Sweitzers.

## Where Did Standard Barns First Appear?

Several first-generation Standard barns of pre-1820 vintage are seen in a few counties in the hearth. Whether they appeared in one county first or spontaneously in several counties is a difficult question to answer.

The 1792 Duportail Standard barn near Devon in Chester County offers a clue. It is a substantial barn, 55 feet long by 37 feet wide with 18-inch-thick stone end walls and rear wall. Obviously, much thought went into the barn's design, and it is unlikely that this was one of the first Standard barns ever constructed. It would seem more plausible that smaller barns of the type were built prior to 1790 that duplicated the dimensions of other area barns, such as Ground barns or (less likely) small three-bay Sweitzers.

The curious thing about the Duportail barn is that it is in a heavily English-settled area. English were settled there as early as the 1680s. Yet its distinctive forebay is based on German traditions. English Lake District (ELD) barns were much more a part of the local cultural environment in Chester County than German forebay barns were. It may be that the Standard barn had its genesis in an ELD barn and the stable wall was recessed several feet to the rear to create the cantilevered effect. In addition, the front wall was changed from stone construction to frame construction. If this scenario is correct, the regular forebay Standard barn form could have been based on English design.

Another possibility is that it is based on a Sweitzer. By eliminating the front appendage or forebay and then moving the stable wall back several feet, the Standard barn type was created.

It seems likely that the Standard barn type first appeared in one or more of a few counties, namely, Bucks, Montgomery, Chester, Lehigh, and Berks Counties. The style may have been deemed so successful that hundreds of farmers wanted to replace their outmoded Sweitzers with this new-fangled barn type as soon as possible.

Beginning in the first quarter of the nineteenth century, farmers had to streamline their operations because of increased commercialization; farm produce was sent to market and profit was paramount. Competition, coming from the Midwest, was the fierce dog of the age, and the Standard barn form was the solution.

## Frontality and Balance

Efficiency was not the only driver of Standard barn design. The Sweitzer barn concept came at a time when Georgian architecture was a major component of the eastern United States cultural landscape. But after 1790, the Federal age ushered in a new architectural tradition, even in the highly conservative German areas of southeast Pennsylvania.

Aesthetics must have played at least a partial role in the popularity of the Standard barn. A manifest incongruity existed between the Sweitzer's steep, almost medieval-like roofline and the front or eave-wall house facades. It was perhaps the symmetrical Georgian and Federal houses with their proportions of balance, uniformity, and frontality that influenced Standard barn design. The short front forebay wall of the Sweitzer barn clashed with the high and proud facades of many German houses. The prominent front wall of the Standard barn eliminated the essentially "hide and seek" appearance of the Sweitzer. Early vernacular houses and barns were often oriented according to weather, terrain, and exposure to sun. But by about 1820 to 1830, cultural tastes had begun to enter the picture. Buildings at that time, including barns, were often proportioned and dimensioned according to an increasingly ordered and regimented culture.

At some point in the first quarter of the nineteenth century, the Sweitzer barn may have even begun to look old-fashioned. It was then that the Standard barn began to prominently play the "modern" cultural fiddle. It did so for more than a century.

## "Modern" Scientific Research

Alternately, one might say that the growing scientific literature on advanced agricultural methods led to the proliferation of larger barns. Newspapers, inexpensive books, and patents in the first few decades of the nineteenth century promoted changes to agricultural practices. In addition, agricultural societies were gaining prominence in the East, such as the Philadelphia Society for Promoting Agriculture, begun in 1785. The society was not intended for the benefit of everyday farmers but, as Stevenson W. Fletcher says in his book, *Pennsylvania Agriculture and Country Life: 1640–1840*, for "wealthy patrons of agriculture who could use their country estates to test new methods developed at home or abroad."

The society performed experiments that influenced the direction of agricultural practices in Pennsylvania for many years, partly by promoting other agricultural societies in the state. While rural working farmers were at first skeptical of "scientific farmers," they were strongly influenced by precedent and "hands on" demonstrations. Agricultural societies proliferated in the ten-year period from 1815 to 1825. The Berkshire (Massachusetts) County Fair, chartered in 1811, set a precedent with a parade, agricultural ball, and other activities, becoming a model of other such organizations, such as the Bucks County Agricultural Society in 1819. By 1825 about fifteen societies were functioning in the state, and many ideas were exchanged. It was a slow process convincing farmers that scientific methods had merit. Up until about 1840, most farmers read only the Bible and *Poor Richard's Almanack*. All this affected how farmers operated and managed the design of their barns.

## Outstanding Standard Barns

### Duportail Barn, 1792

The earliest known Standard barn is the dated 1792 Duportail barn at the old Cressbrook Farm in the Chesterbrook area of Tredyffrin Township, Chester County. Ball (1974) states that the barn was known as "The Federal barn." In the book, a nice oil painting by area artist Barclay Rubicam illustrates the barn. It was actually comprised of a 1792 section and a stone-to-peak bank barn section that included conical stone piers at its front. The section with the piers was dismantled a few decades ago, leaving only the 1792 section. In Ball's book no mention is made that this section is a Standard barn. By pure good fortune, I found this barn in January 1995. It has been recognized as the only eighteenth-century Standard barn ever identified.

The barn is a stone-to-peak structure and is 55 feet, 3 inches long, by 37 feet wide. The barn has three square (2 feet by 2 feet) stone supports below the front forebay wall. A *peiler eck* also occupies each front corner. The rear wall and one end wall, which includes three doors (two at the basement level and one at the upper level), have splayed loopholes. Strangely, the opposite end wall has no loopholes. A substantial stone wall delineates the front barnyard.

Unfortunately, the barn interior could not be examined, but it has apparently been much altered through the years. Other than this, the barn is in good original condition.

If original timbers remain in the barn, they should be dendro-dated to see if the test results actually match the date on the date-stone on the one gable wall. We would then know that the basic form of the Standard barn was actually established at least as early as the last decade of the eighteenth century. The next task is to find out as much history of the Duportail homestead as possible to unravel some of the mystery of its early date.

### DeLong Barn, 1804

The DeLong barn is the second earliest Standard barn ever identified. It was on the east side of Bowers Road just south of Fleetwood Road in the area of Bowers in Maxatawny Township, Berks County. This magnificent barn was demolished due to a tornado that hit the area in May 1998. I first visited it with Robert Ensminger in May 1991. The date-stone was apparently saved.

A stone-to-peak structure with four-bay construction, it measured 106 feet long by 40-½ feet wide in its original mode. This was an unusually long barn for that area of Berks County. It had an atypical stone section several feet wide between the adjacent wagon bays, where this rear wall area was stone. This may have constituted a narrow bay between the wagon bays. Below the long approach rear ramp was an extensive root cellar. The frame wagon doors were original and had surprisingly large vertical hewn pivot beams.

This is the earliest Standard barn ever found with a *liegender Stuhl* roof type. More than a dozen other Standard barns have this roof format. A large granary room occupied the front of the barn on one side.

### Wagner Barn, 1849

The large stone-to-peak Wagner Standard barn, in the Hellertown area of Northampton County, has an overabundance of outstanding traits.

The barn is 80 feet long and 40 feet wide, and its exterior stonework is superior. The best illustration of this is the nearly perfectly plumb rear wall; this is particularly impressive since the barn is so long. Many

The DeLong stone-to-peak barn dated 1804 is the second oldest Standard barn ever identified. Destroyed in mid-1998, it had a German roof structure. (Courtesy of Robert Ensminger)

barns have concave rear walls, resulting from loaded hay wagons placing inward pressure on the stone foundation walls. The front forebay wall retains most of its original vertical siding along with most of its seven original stable wall doors. The door hoods of the end wall basement entries have finely finished front faces with scalloped bottom edges.

The cavernous interior is every bit as spectacular as the exterior, perhaps more so. It has a Principal Rafter System with five Principal Rafter pairs. Most barns of this date have Common Rafter Systems. Three transverse bents consist of timbers nearly perfectly formed and joined with a high level of craftsmanship. Even the boards of the mow-stead walls are joined in the most careful manner. This also applies to the wood elements of the granary at the barn front. The bents have several areas (at the upper girt to post junctures) where double "triple-loop" figures are marked in red pencil.

The basement's wood elements also show superb craftsmanship. Below the one wagon bay are four summer beams that must have been intended for a particular use. Even the full barn length summers have large, unique red pencil marks labeling numbers one through four at the ends of the middle sections. One horse stall has a rare partial plank floor.

The Wagner barn is without precedent on the Pennsylvania German cultural landscape. The only barn that challenges its lofty status is an apparent sister barn named Wagner a quarter mile away, dated 1851. Its level of craftsmanship is just a shade below that of the 1849 Wagner barn.

### Wassergass Two-bay Barn

Most of us are aware of the old saying, "good things come in small packages." Never in the Standard barn world does this old saw apply more poignantly than to a small structure in the Wassergass section of Lower Saucon Township, Northampton County. Until you open the wagon doors, there is no clue to the prize that awaits inside.

The frame Standard barn in Lower Saucon is just a two-bay structure, but its timbering includes its "overkill" Principal Rafter System, seen in far larger barns. A master builder fashioned these timbers, no doubt.

This frame barn is only 24 feet on each side. Its diminutive size is dictated by its two-bay construction. Each front corner has a *peiler eck*. The barn retains all three original stable wall split doors. Remarkably, this almost tiny barn has a Principal Rafter System—an example of over-engineering at its best.

No simple reason exists for this; a Common Rafter System even without any purlin plate system would have been more than sufficient. One gets the impression that the builder knew of no other roof support type. Other far larger Standard barns have Common Rafter Systems.

The barn also possesses bent posts that are normally reserved for much bigger barns. In addition, the *oberdenn* retains most or all of its original overhead boards; several are nearly 2½ feet wide, a rarity.

## Summary

More Standard barns dot Pennsylvania landscapes than any other barn type. The Sweitzer, while a far more aesthetically pleasing structure and a success for several decades (1750 to 1825), was outmoded after the Industrial Revolution got underway. The Standard barn was the answer to a new cultural movement.

The side wall entrance was so efficient that its basic format was never fundamentally improved upon in the nineteenth century. This was also the case in much of the eastern river valleys in New York state and in west-central New Jersey, where the classic three-aisle Dutch-American barn gave way to the side-wall-entry swing beam barn.

One of the best indications of the Standard barn's success is the tremendous number still found almost everywhere in the hearth. Unlike so many other barns, they offer tremendous opportunity to understand what nineteenth-century farmers thought and how they lived their lives.

# Chapter Seven.
# Up-Country Posted Forebay Barns

The Newcomer barn near Coopersburg in Lehigh County is a prime example of the Up-country Posted barn type. The front shed, like so many structures of the type, was built around 1900.

The Up-country Posted forebay barn (UCP) is a subtype of the Extended Pennsylvania barn known as Type A-2 under type A—Extended Supported Forebay barn. These barns date from 1825 to 1920 and many were built in the mid-nineteenth century. In the hearth area, however, the barn type seems to date from the 1875 to 1920 era. They are commonly seen in all of Lehigh County, much of Northampton County, northern Montgomery County, and eastern Berks County. They are scattered about in the remaining hearth areas north of Philadelphia.

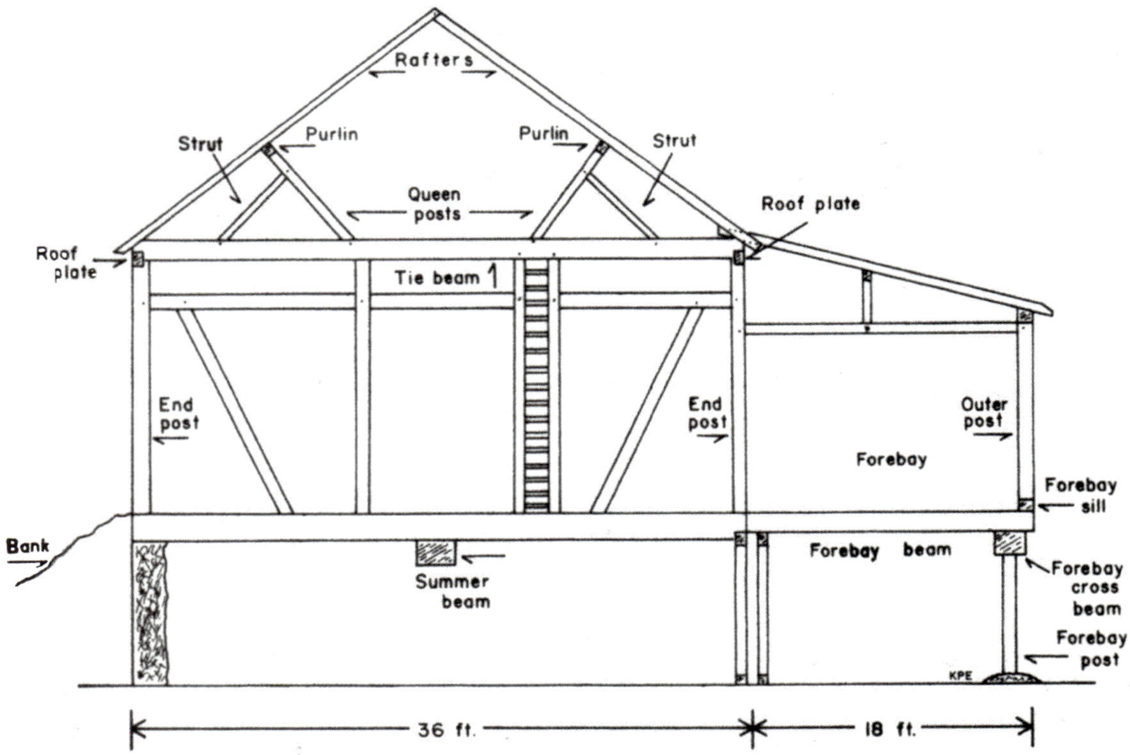


The Up-country Posted barn has two sections—the rear or main section (left) and the front straw shed (right). The main section has all normal framing and the front shed has a framing structure at second floor level, but a totally open area at the level even with the basement of the main section. (Courtesy of Ken Easterday)

UCP barns were usually created when front straw sheds were added to existing closed Standard barns. It did not matter whether the original Standard barns were stone-to-peak, stone-to-eaves-level, or frame. When these additions were built, the existing barns were transformed into UCP barns.

## Construction Aspects of Front Straw Sheds

Front straw sheds were frame structures attached to the front eave walls of existing Standard barns and run the full length of the barns. The sheds often had roof slopes pitched at different angles than the existing roofs. In contrast, Sweitzers almost always have their front forebays pitched at the same angle as the main roof. Recall that the front appendages are integral parts of these barns.

Widths of front sheds range from 8 feet to about 16 feet. Average width of front sheds is 10 to 12 feet.

The front shed extends from the top to the bottom of the forebay wall. The area below the shed floor is open. The front wall of the shed is often supported by a series of round metal columns 5 to 10 inches in diameter. In many cases, the metal columns are original. Often, these columns stand on stone blocks. In other instances, well-formed timber beams are used as posts to support the shed front walls. Often, these posts are rather widely chamfered at their corners. Sometimes the bottom few feet of the beams have deteriorated so that new wood replaced the rotten wood. Other wood posts have been fully replaced, especially those at the ends or front corners, as these were exposed to rain and snow from two sides.

Several windows were often included on the front walls of the sheds. In addition, winnowing doors were always included at the center of the walls in line with wagon bays. Occasionally, with barns that have four bays, two of these doors, equally spaced, were part of the front walls.

Ceilings of front sheds (seen from the ground level overhead as transverse ceiling beams) consist of a number of joists or front-to-back timbers that are either hewn,

A stone Standard barn near Coopersburg in Lehigh County has an added front straw shed supported by original steel supports. In certain other barns, wood posts, some original and often chamfered at their corners, support straw sheds.

A barn in Longswamp in Berks County illustrates the "doubling up" effect of joists of both a Standard barn and a front straw shed. When this occurs, proof exists that an attached shed is not original to a barn site.

milled, or round poles. These joists extend from the bottom of the forebay wall to the front of the shed. Ceiling heights of the joists are generally about 8 to 9 feet but occasionally more.

Interiors are spacious. They sometimes have partition walls but are often one large open space, end wall to end wall. Occasionally, granaries are seen within the sheds, and when present they are always adjacent to an end bay and next to an end wall. In these cases, granaries were often removed from the original Standard barn sections and re-located forward to the front shed. In other instances, granaries were enlarged from their original dimensions (in the Standard barn section) to occupy an area in both the Standard barn section and the shed. In one example in an eastern Berks County barn, a large granary that occupies both the main barn section and the front shed has a central passageway to grain bins at either side.

Most of the time, timbers are connected by mortise and tenon and pegged joinery. Some sheds have hewn timbers and in other cases, beams are milled. Roof rafters are nearly always milled. The milled timbers often have circular saw marks that indicate post-1875 construction.

In a number of situations, when sheds were added to Standard barns, the original vertical siding on the front forebay walls survive. The siding may feature faded barn stars. The siding was often removed, ostensibly to open the passage between the main section and the shed.

It is easy to discern whether a front straw shed was original. When a straw shed was added, the inner ends of the ceiling joists were placed at the top of the stable wall of the Standard barn section. Usually , joists were placed immediately adjacent to the outer ends of the joists of the main section. Thus, for 4 or 5 feet, the joists of both barn sections have a "doubled up" appearance. The outer ends of the joists (end grains) of the Standard barn section are often plainly seen. This condition shows that the shed is not original. In the rare situations when straw sheds are original, the inner ends of the joists are positioned from the stable wall of the main section as "single timbers" out to the front wall of the shed.

Front straw sheds were added to increase storage capacity. After about 1870, the need in the up-country—the areas north, northeast, and west of Chester County—for increased hay and straw storage was acute. UCP barns were the result of new pressures that farmers were experiencing after about 1875 into the first two decades of the twentieth century.

## Three-Bay Versus Four-Bay Constructions

At many historic homesteads, Standard barns were most likely three-bay construction and were converted into UCP barns. Many four-bay barns (and some five-bay barns) also experienced such transformations. One study of UCP barns in Longswamp Township in Berks County revealed that ten barns of the type had three bays. Only two had four bays. Storage was probably adequate in four-bay barns and no straw sheds were needed. Indeed, an interesting statistical study could be done of all the different sized Standard barns in southeast Pennsylvania that have front straw sheds, and determine the percentage of three-bay barns versus four-bay barns that had sheds.

## Examples of UCP Forebay Barns

Many dozens of UCP barns have been recorded to varying degrees. The barns themselves number several hundred to 1,000 or more in the up-country. Their dimensions vary considerably, but their construction methods are similar.

### Laubach Barn (Original)

A rare original UCP barn is at the Laubach homestead off Route 412 in Lower Saucon Township, Northampton County. The main Standard barn section at the rear and the front shed section were built at the same time. The main section is a stone-to-eaves-level Standard barn 40 feet long by 30½ feet wide. The frame shed is 10 feet wide.

Three aspects indicate that the two sections were built at the same time. No "exterior" vertical boards were ever attached to the front wall of the Standard barn section. No granary ever existed in the main section, as there was a granary in the shed. And the stone stable wall was never recessed back from the front stone corners of the main section. From the beginning, the Laubach family intended to build a barn with two sections.

Built circa 1875, it may have planted the seed for other area builders to add straw sheds to existing Standard barns. Original UCPs with stone main sections are rare.

The Laubach barn, circa 1875, in Lower Saucon in Northampton County is an original condition Up-country Posted Extended barn. No doubling-up effect of joists is in evidence. The stable wall was never moved.

The circa 1900 Royce barn in Lower Saucon is an original condition Up-country Posted structure. The main section has a gambrel roof.

## Western Lower Saucon Township Barn (Original)

Another original UCP barn, all frame, is in western Lower Saucon Township about 2½ miles west of the Laubach barn. This barn is a circa 1900 structure with a later style gambrel roof. Unlike the Laubach barn, the rear Standard section has a fully intact forebay area. That is, a recessed stable wall duplicates a Standard barn style. All the upper floor bent timbers are milled, indicative of a late structure.

Both barn sections were constructed at the same time as indicated by three features. The original exterior siding is consistent throughout the entire structure, on all of its walls. No siding ever existed at the front wall of the Standard barn section. The joists in the basements in both sections have a consistent appearance. No doubling-up effect is seen in the front shed.

The Royce barn may be unusual for an original-condition structure where the normally seen *peiler ecke* exist at the front corners of the main section.

The two-section barn in Longswamp has a non-original front straw shed. The doubling-up effect is seen across the full length of this four-bay structure.

The circa 1820 stone Standard barn north of Topton in Berks County has a front shed that was possibly built in 1916, as indicated on a wood plaque on the end wall.

### Red Lion Area Barn (Non-Original)

A medium-sized frame, two-section closed Standard barn with a later-built straw shed is in Longswamp Township in Berks County. The Standard barn section consists of two two-bay sections. The added shed stretches across the full length of the barn. A barn star with the name Wetzel and date of 1848 adorns the rear wall on the two-bay frame addition.

The front shed is 12 feet wide and seems to have been built in one building campaign, likely in the 1890 to 1910 era. One source says that the famed Pennsylvania folklore scholar, Alfred Shoemaker, may have visited this homestead, perhaps in the first half of the 1950s.

### North of Topton Barn (Non-Original)

A large barn with a later-constructed front straw shed is north of Topton in Longswamp Township. The main section is a four-bay stone-to-peak Standard barn. Just below the roof peak of the Standard barn section on one end wall is a wood plaque with the dates 1916 and 1820, probably the construction dates of the two building sections.

The front shed is 14 feet wide, which pushes the upper limit of shed widths. Strangely, the end wall of the shed on the road side (not original) has an enclosed area at the bottom. The 9-foot, 9-inch ceiling is one of the tallest recorded. Often this height is about 8 feet.

## Summary

There are hundreds of Standard barns with added front sheds in the up-country. Likely, many farmers copied their neighbors. Agricultural journals helped disseminate the idea of adding front straw sheds to Standard barns. This basic barn construction idea extended far beyond the hearth into Maryland and Virginia, especially the Shenandoah Valley and upper Midwest. The single-level construction was economical to build, and we may see more of these barns down country roads.

Since initial publication in July 2017, two more original condition UCP Barns were found, both in Albany Township in Berks County.

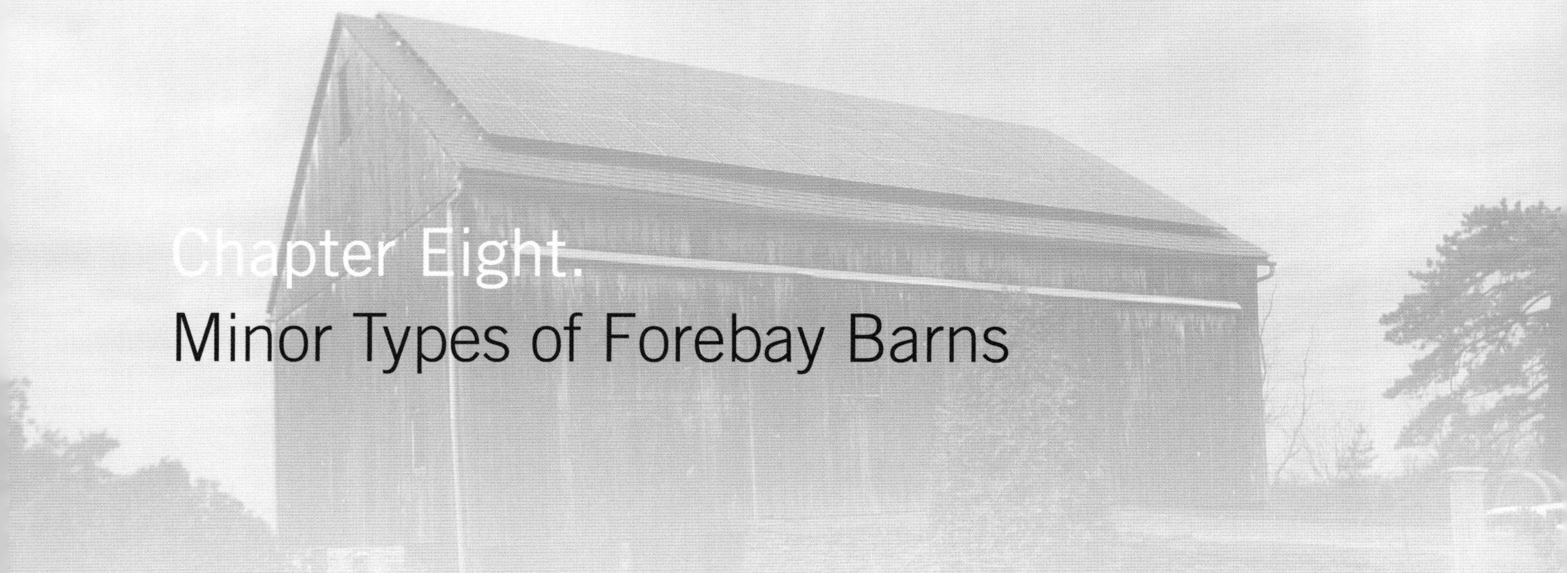

# Chapter Eight.
# Minor Types of Forebay Barns

Five barn styles occupy minor positions in the hearth's barn population. They include the following: stone-arched, ramp-into-gable ends, Double Deckers, Chester County stone-posted, and bank-into-forebay barns. The types listed here are some of the most interesting ones on the landscape. Their basic configurations and structural details deviate considerably from the major barn styles. There are other, less culturally important barn types not included in this chapter. They are front shed or three-gable barns, multiple overhang barns, posted-forebay barns, basement drive-through barns, and rear extension barns. However, a few observations about these barns are worth mentioning.

Rear extension barns are likely the most prevalent of the omitted barn types. They occur in several counties, especially Lancaster County and occasionally in the Lehigh Valley. When barns with these out-sheds are attached to the rear of Standard barns, they create asymmetrical rooflines. The sheds can be either integral to the barns or additions. Some are stone, others are brick, and the latest are frame. Some barns, such as the superb Gockley Sweitzer in Lancaster County, have both an earlier stone out-shed and a later brick out-shed. Usually the rear out-sheds include granary areas.

Front shed or three-gable barns are uncommon in most areas of the hearth. One interesting specimen, however, is in Heidelberg Township in Lehigh County. The frame front gable shed has a forebay. This may be a unique barn type in the hearth area.

Few multiple overhang barns are in the hearth. One is in Lancaster County not far from Lancaster City. In addition, few basement drive-through barns are located in the hearth. One exception is a circa 1820 stone Standard barn in Lower Saucon Township in Northampton County. Likely the best-known example is a splendid stone transitional Sweitzer in Buffalo Springs, Lebanon County. The barn is massive at nearly 130 feet long. Posted-forebay barns are seen occasionally in the hearth including the Lehigh Valley. The Hendrickson barn in Alsace Township, Berks County, is a large example.

These barn types, excluding the out-shed type, may appear in clusters in any county in the hearth. These barns, if they do exist, simply have not yet been identified, as the hearth area is far from being fully surveyed.

## 1. Bank-into-Forebay Standard Barns

The rarest of all Standard barn types is the bank-into-forebay barn. The earliest bank-into-forebay barns were likely built about 1840 to 1850. Most of the barns are frame, appeared after 1850, and were built until the early 1880s. They mostly appear in upper Bucks County along Route 113. Eight such barns have been identified, two of which no longer exist. Several others were likely erected in the area but were never identified or examined. In addition, one other barn of this type is in the Sugartown area of Willistown Township in Chester County. Another exists near State College in Centre County. Most or all seem to have been built after about 1850. One barn is dated 1881.

The genesis of this unusual and essentially localized barn type can't be easily explained. It may be that one farmer in the area near Route 113 experimented with a new-fangled way of getting his hay wagons onto the upper floor at, amazingly enough, the front of his barn. Perhaps this was an accommodation that related to certain topographical aspects of the landscape. It worked and the idea spread to neighboring areas, but apparently not to a great extent. In the immediate Route 113 area, only seven miles separate the barns that are most distant from each other. It is possible that only a few builders erected these special barns.

### Defining Characteristics

- Bank-into-forebay structures have almost all the Standard barn features, including stone foundation walls with wood siding above. A barn off Route 412 in Nockamixon Township, Bucks County, has stone end walls. Even *peiler ecke* are present at the front barn corners. In addition, the internal framing, basement stabling areas, and, surprisingly, stable walls with their split doors are typical of these Standard barns.

- The area where loaded hay wagons entered the barns on the front eave walls (locations of forebays) consists of two aspects that distinguish these special barns from other forebay barns. The first section is the earthen ramp that varies in length from about 20 feet to as much as 60 feet. The ramps may be as wide as 30 feet. They are angled with their widest ends on the side away from the barn. The ramps are angled as accommodations for wagons as they approached the forebay sides of the barns. The ramps have vertical stone walls at their inner ends that vary from about 6½ feet to 7 feet in height. One barn is exceptional in that it has a 9½-foot-high stone wall. Another barn has a stone wall nearly 30 feet long. The walls are parallel to the stable walls. The distances of these walls to the forebay walls vary from 7 to 10 feet.

  The second aspect that occurs in just two barns is the frame ramp shed or roofed overbridge that spans the gap separating the ramp stone wall and front forebay wall. One example that seems authentic is the barn off Route 113 in Bedminster Township, where the shed is 11 feet long by 5½ feet deep. The other barn is off Diamond Road in Hilltown Township, with a shed 12 feet long by 9½ feet deep. Another such barn in Elephant survived until about 2010.

The Hilltown Township barn has a prominent stone ledge or lip at the top of the stone-walled approach ramp near the stable wall. The ceiling joists rest on the ledge.

- All barns but two originally had three bays. A large four-bay barn is a stone-wall-ended structure about one mile south of Silverdale in Hilltown Township, while the other is a two-bay frame structure in Willis Township in Chester County with a one-bay frame addition.

### Dimensions

Six of the eight barns in Bucks County have been measured. The smallest three-bay barn by far is 31½ feet long by 31 feet wide. The largest barn (four bays) is 60 feet long by 40 feet wide. The average length is 42½ feet and the average width is 33½ feet.

A two-bay barn is just 30 feet by 30 feet. The Sugartown barn is a two-section, three-bay frame building where the original section is two bays with a one-bay addition. Not only are the ramp-into-forebay barns the rarest of all Standard barn types, they are also the smallest on average. It seems most likely that the barns occupied relatively small farms and reflect the fact that farmers wanted to take advantage of as much land as possible.

A very large bank-into-forebay frame barn is in Hilltown Township in Bucks County. It has the widest approach ramp of any of the barns.

An upper Bucks County stone end-walled barn had its roof structure removed on one end bay that resulted in a lean-to section on the right side.

This Sugartown bank-into-forebay barn in Willistown, Chester County, is a two-section structure. This is a museum building. The barn is of two sections: a two-bay original part and a one-bay addition.

### Roof Structures

All the barns of the type have Common Rafter Systems. Since the earliest built ones were constructed after 1840 or 1850, they are, in general, too late to have the English based Principal Rafter Systems and certainly the German style *liegender Stuhl* roof system incorporated into them.

### Rear Walls

As expected, the rear walls of the frame barns, where normally wagon doors are located, are solid fully-frame constructed, from ground to the eaves. The one stone wall ended barn has a solid stone rear wall. Wagons entered these barns through front wall wagon doors.

### Two Interesting Bank-into-forebay Barns

Four-bay barn. One unique bank-into-forebay Standard barn resides down an old farm lane off Fairhill Road in Hilltown Township in Bucks County. At 60 feet long and 40 feet wide, this imposing frame structure is the largest of all barns of the type. The earthen approach ramp is itself 60 by 30 feet, larger than several of these barns.

The rear stonewall of the ramp is 9½ feet high, and an unusual ledge of stone at the top of the wall projects several inches beyond the rest of the wall below. This wall is about ten feet to the front of the stable wall. The stable wall retains all five of its original split doors. This is the only four-bay constructed barn of its type (except one original two-bay barn). Canted queen posts support the purlin plates. In an unusual treatment, built-in ladders are all off-center on the bents. A carved date on a stone in the stable wall – 1881 – may be the time of construction.

Two-section barn. One barn is 35 miles southwest of the epicenter of the bank-into-forebay barn building area. This much "displaced" barn of frame construction is the only two-section barn of the type known.
The original part is of two bays built circa 1860 and the one-bay frame addition was built a decade or two later. The original section must have supported a small farm. The farm size might be explained by the fact that the barn is quite close to the intersection of two major roads. Large farms would infrequently have their barns at such locations.

A several-foot-wide passageway always exists between the stable wall and the rear stonewall of the earth ramp. This barn is now a museum place in the area of Sugartown in Chester County.

## 2. Double Decker Barns

Double Deckers, wherever they may occur, dominate agricultural landscapes by virtue of their sheer size. Their great dimensions lend a sense of cultural pre-eminence at homesteads where farmers had many acres of tillable land. Owners of such farms and barns were in for the long haul and any expenditures reflected their dedication to the tasks at hand. Double Decker owners were not amateurs; they were businesspeople every step of the way.

Recall from an earlier discussion that across almost all rural landscapes in the east, three major forms of barns are recognized by barn observers and scholars alike—one-level barns (ground type), two-level barns (banked structures) and three-level barns or Double Deckers.

Double Deckers have long been recognized in barn literature. Images of these barns are seen in the books of Shoemaker, Dornbusch-Heyl (Type L), Ball, and Ensminger (see appendix). Berenice Ball gave the Double Decker extensive coverage in her book on Chester County barns, with illustrations of at least eighteen Double Decker barns.

Double Deckers are seen in many areas of principally English-settled (Quaker) areas of southeast Pennsylvania, as well as other states. A Double Decker is known in

The Adam Geringer barn dated 1836 in Bethlehem Township, Northampton County, is a museum structure. It has lost much of its original interior features.

the heavily settled Holland-Dutch area of Ulster County in eastern New York state. Robert Ensminger has reported three Double Deckers scattered in the western Catskills in New York. Another was found south of Elmira, New York. One Double Decker is located in southern Wisconsin in Greene County, and another has been sighted in Benton County, Oregon, near Eugene. Likely, many other Double Deckers occur throughout the US. Abroad, the Double Decker form is quite often seen in Alpine-Germanic Europe (Germany/Switzerland/Austria). Only one in England (in Dacre) has been identified, by Ensminger.

Double Deckers are found in most counties in the hearth area in Pennsylvania, including Chester (9), Lancaster (6), Delaware (3), Berks (2), and Northampton and Montgomery (1). Many more have been viewed from a distance, especially in Chester County. None have yet been identified in Bucks, Lebanon, or Dauphin Counties. Far more Double Deckers will certainly be identified, particularly in Chester and Delaware Counties, Bucks County's southern half, and perhaps in Montgomery County.

## Defining Characteristics

Double Deckers include the following features:

- It is a three-level structure that includes an upper floor for storing crops in mows and in areas high above the wagon bays. The middle floor has sink mows that vary in depth from

Baldwin's Book Barn is a large Double Decker in West Chester, Chester County, dated 1822. It has much of its interior structure including the superb basement level end wall doors.

6½ feet to 11 feet, and a grain compartment, usually in the middle of the floor between the sink mows. A basement is used for stabling animals. Few basements retain their original staked mangers and manger troughs.

- It is always a banked structure. The bank or approach ramp typically appears at the rear of the barn where loaded hay wagons traversed. In a few special cases, such as the spectacular 1795 Brechbil barn near Paradise in Lancaster County, the wagon entries and their long earthen ramps are located at each end wall. The wagons entered the barn through 11- to 12-foot-high doors.

- The front eave wall is often stone and is punctuated with wall openings. Some barns have framing above the stone foundation walls and date from the post-1830 or 1840 era.

A great-sized sink mow is in a large Double Decker in Montgomery County, Douglas Township, near the Berks County line. The roof structure (not seen) was dramatically altered.

The 1806 dated Pennock Double Decker in Chester County, West Marlborough Township, doubles as a "Standard barn" with a middle floor level with sink mows and grain compartment area.

- In many barns, stable door entries are on the basement level at the front. Some have a barn-length pent roof that extends about 4 feet. Original pent roofs are quite rare. As in English Lake District barns, the pent roofs protected the stable doors and farmers from the weather.

- A few of these barns were originally built with forebays or cantilevers at the barn front. An excellent example is the spectacular Pennock Double Decker "Standard barn" dated 1806 in West Marlborough Township in Chester County. In addition, at least two Double Decker Sweitzers are in Lancaster County—one in the Conestoga area and one, immensely tall, near Marticville off Route 324.

- Many Double Deckers have prominent bridges at their rear eave walls. The front walls of the earthen ramps that lead up to the barn bridges are often separated from the rear walls by 8 to 12 feet. The bridges that span the separation or gap have plank floors that often had to be replaced due to exposure to the weather. Some of the floors of the bridges were originally protected by gable-ended structures.

The Pennock barn has an access door into the middle floor level on its rear wall. Such door positioning is normal.

Bridges are in line with the tops of the middle floor level. Many of the barns have rear wall door entries below the bridge that lead to the middle floor level. One barn in Chester County has a bridge that includes corncrib areas at each side. Another barn in Willistown Township in Chester County has a spectacular bridge.

- In a certain way, Double Deckers with straight stone front walls are quite similar to English Lake District (ELD) barns. The fundamental difference is the inclusion of a middle floor level in Double Deckers. We may regard Double Deckers as vertical extensions of ELD barns. Double Deckers, except the middle floor level, were similar to other bank barn styles in Pennsylvania in terms of interior timber framing and masonry construction methods.

## Age Distribution

The earliest known Double Decker on record is the dated 1795 Brechbil barn near Paradise in Lancaster County, in what was a Quaker settlement surrounded by Germanic settlements. During the first two decades of the nineteenth century, Double Deckers were probably being erected in fair numbers. Witness the 1806 Pennock Standard forebay barn in West Marlborough Township and the dated 1809 Wickersham barn in Newlin Township, both in Chester County. In the following decade, a dated 1817 barn was built in Salisbury Township in Lancaster County. By this era, the Double Decker barn form was proliferating except in the Up-country.

One apparent early Double Decker is seen in Margaret Berwind Schiffer's book *Survey of Chester County, Pennsylvania, Architecture: 17th, 18th, and 19th Centuries.* This stone-to-peak barn, in West Bradford Township in Chester County, supposedly had three levels—"stables and yard on the lower level, threshing floor on the second level, and mows on the third level." Threshing floors, however, are normally on the top floor. The barn was dated 1804 and was 46 feet by 44 feet, plus a frame straw shed was supported by stone piers.

The Double Decker form likely started to see its demise by the Civil War era. The latest known barn is the Davis barn in West Vincent in Chester County dated 1869, a large stone-to-peak structure. It seems likely that a few other barns were built into the 1870s.

## Dimensions

The range of dimensions of Double Deckers exceeds those of English Lake District barns. Twenty-three Double Deckers have been identified in the hearth area, fourteen of which have been measured. Lengths range from 46 feet to 100 feet. Widths range from 35 feet to 56 feet. Four of the barns have lengths of more than 70 feet. Most of the barns are 50 feet to 65 feet. The average length of Double Deckers is 63 feet and average width is 44 feet. The average barn length is not unusual but the average barn width is distinctly greater than most if not all other barn types. The Hugh Hodge barn built in the 1830s in Centre County (outside the hearth area) is a colossal 119 feet long by 60 feet wide.

## Roof Structure

Just a few Double Deckers were built with Principal Rafter Systems. They include the Brechbil barn and the rare and large frame Double Decker Sweitzer barn in the Conestoga area of Lancaster County. Others likely have the roof type. The latter barn is unique in that the bottom ends of the Principal Rafters join the major wall posts near their top ends instead of at the top of the upper ties. Most other Double Deckers have Common Rafter Systems.

## Outstanding Double Decker Barns

1809 Wickersham Barn in Chester County. One of the most outstanding barns of any style or type in

The 1809 Wickersham Double Decker in Newlin Township, Chester County, began as an English Lake District barn built in 1769. It has a rare king post extending from below the rare swing beam to the roof peak.

Pennsylvania is the Wickersham barn, originally an English Lake District barn built in 1769. Its stone end walls (and possibly part of its front wall) were expanded in the construction of a larger Double Decker in 1809. A gabled-frame barn bridge was added to the rear. This barn, part of an active farm, is in Newlin Township in Chester County.

The interior has spectacular timber framing, including a rare and prominent king post centered above the wide wagon floor and a swing beam with which it intersects. Only two other extant barns with king posts are known in Pennsylvania. King posts are stout vertical posts that usually emanate from an upper tie beam at the top of a bent to the roof peak. In the Wickersham barn, however, the king post springs from the swing beam lower on the bent. Swing beams outside of central Bucks County are rare in southeast Pennsylvania. The son of the owner of the Wickersham homestead claims that "gold" built this barn.

Brechbil Double Decker in Lancaster County. The Brechbil Double Decker is the creme de la crème of agricultural buildings in the hearth area due to its sheer magnificence and unique aspects. It is just south of Route 30 in Paradise Township in Lancaster County.

The exterior dimensions are 65 feet long by 56 feet wide. Although not overwhelmingly large, its width is almost unprecedented. It is a stone-to-peak structure with splayed loopholes on its exterior walls. The upper ends of the gable walls terminate in a parapet-like refinement. Parapets are end wall extensions of 10 to 12 inches above the regular level of the roof that ostensibly protect the roof from wind damage. The presence of such a parapet roof on a barn is unique in North America.

Two of the barn's most intriguing aspects are the prominent stone arches over each end wall wagon entry. Loaded hay wagons traversed the approximately 80-foot-long stone-walled east approach ramp and entered the barn where their contents were unloaded. The wagon

The dated 1795 Brechbil stone Double Decker near Lancaster City has several unique aspects including a parapet roof or end wall edge extension, and an interior structural element called a scissors truss, among other features.

then exited the opposite end wall. The west ramp is about 100 feet long. The two earth ramps alone make this a unique site.

The original date-stone includes the Brechbil family name plus the construction date of 1795. This date post-dates the house by two years.

The barn interior is cavernous and has a Principal Rafter System. Seven Principal Rafter pairs plus a series of common rafters and an unusual triple-tiered purlin plate system define the full roof support. In addition, it contains a rare series of scissor trusses. Four of these trusses were originally included in the roof structure but only three remain. Each truss consists of two "blades" that cross each other; each blade extends from near the wall plate of each eave wall to a point about two-thirds up each Principal Rafter in a pair. The function of the trusses is likely to hold or support the long Principals in their original positions. Just one other barn in the East is known to have scissor trusses—a late eighteenth-century stone Ground barn at the Hermitage in Virginia.

Both middle and upper floors have a rare cruciform plan. That is, an 11-foot end-to-end central aisle crosses

The Brechbil stone Double Decker retains its original date-stone of 1795 high on one gable wall.

an 18-foot-wide side-wall-to-side-wall transverse aisle that forms a cross that creates the cruciform plan. The cross aisles on the upper floor level facilitated gravity-fed farm crops into sink mows on the middle floor level. One interesting aspect of the upper level is that, because

A rear view of the Pennock Double Decker in Chester County reveals a long approach ramp plus the wood ramp next to the stone rear wall.

of the aisles, fully constructed transverse bents are not present. Therefore, no bay divisions are seen on this floor level, a rarity in a timber framed barn.

The middle floor consists of two characteristic entities. The first is the presence of four sink mows, positioned at each square created on the floor by the two aisles. The sink mows are 11 feet high compared to the typical 8 feet. The other characteristic feature is the grain compartment bins that have not survived in the barn. They are usually to the sides of the sink mows.

In the basement, only the ceiling timbers remain; staked mangers and manger troughs have been removed. May this marvelous barn last another 220 years.

Pennock Double Decker in Chester County. A rare Double Decker "Standard barn" is in West Marlborough in Chester County, on what was a Quaker homestead. This is another large barn 61 feet long by 40½ feet wide. Few other such Double Decker "Standard barns" exist in the hearth area.

The Pennock barn has some of the finest carvings on its one original mow-stead wall, including one of the very earliest dates of 1812.

Opposite Page: The Wilkinson barn has had two barn lives, the first as an English Lake District barn dated 1777, and the second as a Double Decker built about 1820 or 1830.

The Pennock barn doubles as one of the oldest Double Deckers and one of the oldest "Standard barns" in the state. This suggests that the Standard barn may have English origins. Recall that the oldest Standard barn and also the only eighteenth-century Standard barn thus far identified, the dated 1792 Duportail stone barn, is found in the heavily English-settled area of Devon in Chester County.

A few curious architectural elements are atypical of a barn built in this era. The roof structure is a Common Rafter System. Many barns of German and English origin of this era have Principal Rafter Systems. The rafters are also milled, whereas most pre-1850 barns have hewn rafters. Another distinct oddity is that some of the rafters appear to be spliced. If the rafters are original, this is extremely rare. In addition, the purlin plates that support the rafters are canted, as are, of course, the canted queen posts that support the plates.

Wilkinson Double Decker Converted from an English Lake District Barn. Another excellent Double Decker started its life as an English Lake District barn. The barn is less than a mile from the intersection of Route 41 and Route 1 in London Grove Township, Chester County. On the front stone wall is the original 1777 date-stone of the first built section with the initials F. H. W. for Francis and Hanna Wilkinson. This is one of the earliest positively identified (former) English Lake District barns known in southeast Pennsylvania. The barn was later converted to a Double Decker, likely in the second quarter of the nineteenth century. Its dimensions, 56 feet long by 36½ feet wide, were maintained.

The roof of the original barn was removed and raised 6½ feet. The wagon bay then became the lower deck of the new barn that include grain bins and sink mows. The original wagon floor with planks 15 to 16 inches wide survives. The floor with original planks may be one of the oldest wagon floors in the state.

One remarkable survival of the original barn is the pair of canted queen posts embedded in each stone end wall. Canted queen posts are normally included in barns after about 1830 or 1840. This is one of the earliest documented uses of these posts in the state.

Hodge Barn in Centre County. Although not in the Pennsylvania hearth area, the Hodge barn near Coburn in Centre County, about fifteen miles east of State

The Wilkinson barn has its original date-stone on its front wall, 1777, with Wilkinson family initials.

College, is an astonishing structure, and is worth mentioning in this discussion of Double Deckers. It is one of the greatest timbered barns I have ever seen in North America. This Double Decker Standard barn built ca. 1850 has frame construction and measures about 120 feet long and 60 feet wide. It may have the greatest volumetric capacity of any vernacular barn in the state.

The end walls and front wall stretch beyond what the eye can see. The great size dictates the rare inclusion of several long braces (1 foot on each side) to support the front forebay wall. The braces extend from the bottom of the stable wall to (or near) the front forebay wall sill.

Inside, the unsuspecting visitor is greeted with a veritable forest of timbers almost beyond belief. A local old-growth virgin pine forest probably supplied the great numbers and sizes of timbers. The roof support type is a Principal Rafter System that prominently features seven Principal Rafter pairs. Part of the roof support is a very rare triple tiered purlin plate system. This three-tiered system was necessitated by the great roof expanse. Some of the posts of the bents are 18 inches across at their top ends. Putting another exclamation mark on the framing are four summer beams in the basement. Another barn with the same number of summers is the overwhelming 150-foot long Kline frame Standard barn in Jackson Township, in Lebanon County.

The master builder of the Hodge Double Decker and his helpers deserved an extended vacation when they completed the construction. Hopefully, they took it.

### Summary

The creation of the Double Decker barn form was a response to changing farm economies in the late eighteenth century. It seems most reasonable that Quakers are responsible for the conception of the barn form in Pennsylvania. The idea disseminated to a number of areas in the hearth area and at least to some extent beyond (Hodge barn). The Double Decker form may be considered revolutionary in that it resulted in a markedly recognizable barn building tradition for perhaps seventy-five to one hundred years. Farmers must have been excited about the ability to expand barns vertically rather than horizontally. If a direct prototype had a European origin it almost certainly did not stem from England. Just one Double Decker, in Dacre in Cumbria, has ever been reported from the mother Quaker country of England. Many Double Deckers were built in Germany, Switzerland, and Austria. Whether the Double Decker notion came from one or more of these countries is not known.

## 3. Chester County Stone-Posted Barns

Amos Long Junior, a Lebanon County historian, commented in *Barns of Chester County, Pennsylvania*, "An appreciative look at the massive eighteenth- and early-nineteenth-century stone barns with their pillar-supported forebay found in Chester County and its environs will reveal that they are among the finest examples of the Pennsylvania barn." Not to diminish Long's view, but this is not likely true. Nevertheless, this distinctive style has long enjoyed iconic stature in Chester County and is recognized by historians.

In the aforementioned book, author Berenice Ball devotes an entire chapter, including twenty-five illustrations, to stone-posted barns.

Other books also mention these barns. Ensminger classifies the barn as one type of the Extended barn. He assigns the date range of construction—1800 to 1870. In the book of Dornbusch and Heyl, it is one barn type they identified. They claim that this barn type is a "further development" of Type H that is the

A stone-posted barn in Willistown Township in Chester County is dated 1819. The barn has additions and may have a record number of stone posts—fourteen, including half posts.

Some Chester County barns have either square or rectangular stone supports in lieu of conical stone supports.

Radley Run Country Club in Chester County has a series of conical stone supports that seem likely to have originated from a barn of the type.

type that Ensminger denotes as the Standard barn. However, Ball firmly states that the stone-posted barn along with several other barn types in Chester County are Quaker (English) based structures. Standard barns, as Ensminger proclaims, are German based. In the book of Arthur and Witney, two excellent views of barns with the stone supports in Chester County are depicted.

## Defining Characteristics

- Barns with stone supports held up the front walls of straw sheds that were joined to existing bank barns or an integral part of two-section barns. Six relocated supports are seen at the Radley Run Country Club in Birmingham Township. At least one support is brick.

- The supports per barn site vary in number from one to five. In some cases, the corners are supported by stone ell-formations akin to *peiler ecke* found in many Standard barns. At other barn sites, additional stone piers support side sheds and increase the number to eight or ten supports. The barn complex with four sections in Willistown Township may have the record

This regular two-level bank barn in Chester County has a stone-post-supported frame front shed that might be called a kind of abbreviated Up-country-like front straw shed.

number of supports, with fourteen. A dramatic example of stone supports is seen in the Arthur and Witney book (1972).

- Most individual stone supports are distinctly tapered. The diameters at their tops are generally 15 to 18 inches and at their bases are 36 to 40 inches (Ball 1974). Their heights average about 8 feet. A few unusual barns have rectangular stone supports.

- Many stone supports are plastered. This accomplished two things: it protected the stone from erosion due to rain and snow and protected farm animals coming in contact with the piers' rough stones.

- Front sheds usually extend across the full length of the main barn sections. The sheds were quite wide, averaging 12 to 15 feet. The front shed at the Willistown Township barn site is 16½ feet wide. The sheds were almost always frame. They have an open area to the immediate front of the stable wall of the main barn basement. The layout of both shed levels closely parallels the front shed conditions of Up-country posted barns.

## Distribution

Most of these barns are seen in Chester County. Before 1940 and the growth of suburban development, hundreds of stone-posted barns dotted the county. Some of these barns may be seen near West Chester, Marshallton, and Mendenhall (Ball 1974).

Delaware County, to the immediate west of Chester County, also has a number of stone-posted barns. A frame bank barn with two stone supports is in the Jarrettown area of Horsham Township off Route 152 in Montgomery County. Another exists in Exeter

One barn outside the hearth in Schuylkill County has three massive stone post supports in the basement. They have great circular support plates at their top ends.

Township in Berks County at a Boone family property. Another is a barn at the Springton Manor Farm in Chester County. Ensminger recalls three barns with stone posts, all in Berks County. One was in the extreme southeast section of Amity Township. Another barn with four stone supports was along Forgedale Road in the eastern part of the county. The last was about one mile south of Lenhartsville west of Route 143. In addition, one barn in Delaware County in Thornton Township has two barn sections, each with a large half-round post adjacent to each other. Together the half posts form one large stone post. Occasionally, barns outside the hearth area possess conical stone posts. One outstanding example is a few miles east of Orwigsburg in central Schuylkill County. Four stone supports in a basement (non-forebay supportive) of a frame addition bear the major weight of an upper level. These tapered supports are far larger than any seen in Chester County. It is difficult to know if any of these barns were influenced by Chester County building traditions.

A barn with three conical exposed brick piers is in Somerset County, New Jersey, on Route 202 near Bernardsville.

## Age Distribution

One of the earliest stone-posted barns on record is the 1806 Williams barn about two miles south of Phoenixville in Charlestown Township, Chester County. Another early barn in Willistown has a date-stone of 1819 that, with its shed, has fourteen stone piers. However, it appears the front shed section is not original. The earliest definite date of a barn with a front shed with piers is at Prospect Hill Farm, at the John Johnson residence in Willistown, Chester County (Ball 1974). Four piers support the shed. Another barn at the home of B. F. Fisher in Schuylkill has a front shed supported by what appears to be at least two stone piers.

Ensminger states that the barn type was first built with front sheds added to the main rear barn section (Ensminger 1992). After farmers saw these barn transformations taking place, they engaged in copy-cat fashion where original condition two-section barns were constructed. Then the barn type essentially became stylized. It seems likely that front sheds with conical piers were built into the last quarter of the nineteenth century.

## Dimensions

The range of exterior dimensions of stone-posted Extended Barns approximates those of, say, English Lake District barns with which they share a common origin. Only two stone-posted barns (main sections without front sheds) were measured in the hearth area. Lengths are 60 feet and 70 feet. Widths are 38 feet and 35½ feet, respectively. It seems that other barns achieve these basic lengths. The front sheds, as mentioned, are about 12 to 15 feet wide.

## Roof Structures

The front sheds have simple milled rafters. With one exception, roof structures of the main barn sections have not been examined. The Williams barn has milled rafters. Barns with Principal Rafters may exist but barns with the early German style roof likely do not exist.

## Two Outstanding Stone-Posted Barns

One of the most outstanding barns of the type is the 1806 Williams-Denenberg barn in Chester County. The barn is set far back on a country road in northern Charlestown Township. It underwent a major renovation in the 1960s that modified both the front shed and main barn.

The main barn section of stone has three bays and measures 60 feet long by 38 feet wide. The barn is

The Williams-Denenberg stone-posted barn is likely the best known barn of its type in any county. It was built in 1806.

stone-to-peak on both end walls. On the house side end wall just below the peak is the carved date, 1806, and initials of homestead family members. In the middle of the front wall, in line with the 16-foot-wide wagon bay, is an opening to the upper level of the front shed.

The main section interior has been modified. Rafters are milled, and canted queen posts support the purlin plates. No granary survives. Almost all of the wagon floor planks are original.

The front shed is 14 feet wide but has been much modified. The original rafters seem to have been replaced by round pole rafters. The front wall has hewn timbers. The ceiling timbers are hewn.

The basement level now contains modern horse stalls. The original two summer beams remain, but the original stable wall doors have been replaced. Interestingly, four joists that over-sail the stable wall are spliced into the joists of the front shed, suggesting that the shed may have been an integral part of the barn.

Four nearly identical tapered conical piers support the shed's front wall. The 8½-foot-high piers vary from about 11 feet to almost 13 feet on center. The far end of the shed has a wall of stone that is not integral to the main barn, as indicated by a vertical seam where the two stone wall sections meet.

Despite these modifications, the structure retains most of its basic design and fabric.

Another outstanding stone-posted barn is in the Sugartown area of Willistown, in Chester County. The three-bay main section is dated 1819 and its basic proportions are similar to the 1806 Williams barn. Its end walls are stone-to-peak. The main stone section is 70 feet long by 35½ feet wide. The shed at the front is 16½ feet wide and also has two frame additions, each with stone piers.

Most of the interior has lost its original framing, including the roof structure. Now the building has a number of large multipurpose rooms.

This stone-posted barn in Willistown, Chester County, has an end wall addition that has several stone posts.

A close-up of one of the half-section stone supports in the addition section of the Sugartown area barn.

The outstanding feature of the four-section barn is the fourteen stone posts or piers. Quite remarkably, incorporated into the original stable wall are five half-section, conical 7-foot piers. It is not known if other barns have these piers integral to the stable walls. If the piers are original, they may be one of the earliest examples of conical piers in the county. In addition, four full piers support the front shed. The first side frame shed has a pier at its front and an inner half-round pier. The outermost side frame shed has four piers, each 8 feet high; these are not conical, indicating that the shed was built several decades after the main section and main front shed.

## 4. Gable Ramp Standard Barns

Another structural type that appeared on the forebay barn scene in the hearth area in the second half of the nineteenth century is the Gable Ramp Standard barn. This barn falls into Type F-1 in the Ensminger classification system (Ensminger 1992). This unusual barn form is similar to bank-into-forebay barns.

While not as rare as the bank-into-forebay barn, several dozen ramp-into-gable wall structures survive in Lehigh, Northampton, Montgomery, Berks, and Lancaster Counties.

The barn's origin is difficult to determine. Perhaps the barn type was an adjustment to the terrain at particular homesteads; the gable wall entry may have saved space or was easier to access on some farms. Whatever its origin, the idea took off, especially in Berks County, but never became widely entrenched in the hearth's lexicon of barn building traditions. Likely, only several builders erected these barns. Gable-ramp house-barns occur in Alpine Europe but no apparent relationship exists between these two building types.

### Defining Characteristics

- Gable-Ramp Standard barns have many regular features of Standard barns. This includes stone foundation walls and exterior wood

Long earth approach ramp leads to the gable wall wagon entry on this gable ramp barn in northern Bucks County.

siding or occasionally stone end walls. *Peiler ecke* are also present at the front barn corners. Certain internal framing, basement stable areas, and also stable walls with their split doors are the same as in regular eave wall forebay Standard barns. The earthen approach ramps at the gable ends average 25 to 30 feet long.

- Several major differences exist, however; the first is that the main wagon entries are on the gable walls. Since loaded wagons have to traverse these particular barns in a longitudinal manner, or end wall to end wall, transverse bents are absent. Therefore, rather than having transverse bays on the upper floor a center longitudinal aisle stretches from gable wall to gable wall.

- The aisles average about 13 feet wide. One large barn has a 16-foot-wide aisle. At the sides of the aisles are mow-stead walls. The aisles are framed by longitudinal timber bents. In

A gable ramp frame barn in Longswamp Township in Berks County has a "rotated" central aisle wagon bay with *oberdenn*. The floor planks run transversely and basement summers and joists run 90 degrees from the norm.

The summer beam in the Longswamp barn receives joists that run both transversely and longitudinally. This summer beam is scarf jointed with an interlocking key.

turn, the wagon floor planks run transversely instead of longitudinally. Above the aisles are transverse *oberdenn* beams, above which farm crops were stored.

- At either side of these central aisles are side aisles for mows. The mows have the same widths as the center aisles. Mows are also arranged longitudinally, or parallel to the eave walls.

- The front side mows include center alcove areas. The alcoves vary in width from 5 to 10 feet and extend several feet in the middle areas of the front aisles. The alcoves have sheathed wood walls with a door to a granary. These granaries have the normal wood ceilings, rear walls, and grain bins.

- The arrangement of basement joists is radically different from most other Standard barns. The beam placement follows the lead of the middle aisle on the upper floor. That is, the basement joists are rotated 90 degrees from the usual layout. Some of these ramp-into-gable barns have double transverse summer beams. In turn, the joists are set longitudinally where their ends join the sides of the summer. But the joists below the front aisle of the upper floor are set transversely. This is the normal appearance in Standard barns. It is these front joists that cantilever over the stable wall out to the front forebay walls. The forebays are 4 to 5 feet wide. Other barns of this same type have the regular longitudinal summer beams, such as the barn in central Longswamp Township, Berks County. Longitudinal joists below the middle aisle accommodate the transversely set wagon floor planks.

- Most of the barns are frame above the stone basement. At least two barns are stone to the eaves level. One unique barn, just north of Lenhartsville in Berks County, is frame with sets of wagon doors at each gable wall. Loaded hay wagons entered one end wall and exited out the other. This barn is one of the largest structures of the barn type.

## Age Distribution

The earliest ramp-into-gable wall Standard barns were stone and likely built 1840 to 1850. The majority of the frame barns were built after 1850 or 1860 until about 1880 to 1890. One frame barn in Lower Saucon Township, Northampton County, is dated 1889.

## Dimensions

All the barns have three aisles. A few of the barns have been measured. They average about 40 feet to 42 feet at their eave walls and about 38 feet to 39 feet at their end walls. The 1889 barn is rather large at 60 feet long and 38 feet wide. Likely the largest barn is the double entry barn mentioned earlier, which could not be measured.

With few exceptions, these are small barns. They compete with bank-into-forebay barns as the smallest of all the Standard barn types and occupied relatively small farms.

## Roof Structures

All the barns of the type have Common Rafter Systems. Since the earliest were built after 1840 or 1850, they arrived too late to have the earlier-style Principal Rafter Systems or the German style *liegender Stuhl* roof structure. The latest dated barn, 1889, has a gambrel style roof with milled rafters.

The Greenwich Township barn in Berks County is one of the largest barns of its type. Perhaps uniquely, both end walls with wagon doors have earth approach ramps.

## Outstanding Gable Ramp Standard Barns

Greenwich Township Barn. This ununusal frame barn in Berks County has two gable wall entries and is quite large, likely 50 feet to 60 feet long.

Longswamp Township Barn. Another excellent gable ramp forebay barn occupies an old farm in central Longswamp Township in Berks County. The barn measures 42 feet long by 40 feet wide. Original end wall wagon doors and much of the vertical siding have been replaced. The split door in the middle of the forebay wall is original.

All five of the split doors (except one lower door half) and three windows of the stable wall are original. A *peiler ecke* is seen at each front corner of the barn. Each alcove area has one or two built-in shelves. The forebay is 4 feet wide.

The interior of the upper floor or loft is nearly one hundred percent original. Such a high level of authenticity is rare in any bank barn in Pennsylvania.

The roof structure consists of sixteen rafter pairs. All rafters are hewn, rather unusual for a post 1860 barn. Later era canted queen posts support the purlin plates. This structure has three longitudinal aisles—a middle aisle and two side aisles. The middle aisle, where hay wagons traversed, is 14 feet wide while each side aisle is 13 feet wide. The aisles are delineated by longitudinal bents. This amounts to four bents, all oriented end wall to end wall. The bents are secured transversely with side wall to side wall ties. Each bent has two long braces, one near each end. Most or all loft timbers are hewn. The structure consists of hardwood timbers including oak and/or chestnut and tulip poplar. Marriage marks appear at timber unions. All original *oberdenn* beams have survived.

Both mow-stead walls are almost all original, including one door on each wall. One door leads to a basement staircase. The wagon floor consists of nearly one hundred percent original floor planks up to 17 inches wide.

In Berks County, the Longswamp gable ramp barn does not retain much of its original exterior envelope. In stark contrast, the upper floor interior is highly original.

The granary of the Longswamp area barn is about centered on the front wall and is mostly original.

A door in the middle of the front side aisle leads to a 5-foot-wide alcove that stretches the full width of the aisle. At the front of the alcove, the ceiling is slanted as a rare accommodation that opens to the front window at the upper half of the forebay wall. The window allows much-needed light into the alcove. All original wood members of the alcove remain, including its original boarded ceiling, boarded side walls, floor, and rafters of the slanted ceiling. An alcove side wall contains an original door that leads to a granary in the aisle.

The granary has a 3½-foot-wide middle passageway with three grain bins at each side and its original wood elements have been preserved. The granary entry door and other wood pieces have many slash marks that recorded grain bushels.

The basement is far less original, although the ceiling beams remain. This includes the two summer beams and all three sections of ceiling joists. Each summer is spliced once. The four joists below the middle aisle are oriented end wall to end wall and, incredibly, each joist consists of one timber, about 40 feet long. Below each side aisle are a series of transversely set joists. The front set of joists cantilever beyond the stable wall to the front forebay wall.

This superbly crafted stone arched barn along Route 611 near Doylestown in Bucks County likely has the best constructed stone walls of the barn type. It has five stone arches.

The middle stable wall door leads to a wide feeding alley nearly 9 feet wide. No regular Standard barn has an alley more than 5 feet wide. At either side of the alley are a cow stable and a horse stable. All original staked mangers and manger troughs were removed decades ago.

### Summary

The gable ramp barn represents the biggest departure from other Standard barns. The orientation of timbers and other wood elements make it structurally different from any other barn type. The circa 1870 Longswamp barn tells the story more poignantly than others of its type.

## 5. Stone Arched Standard Barns

This expression of the Standard forebay barn appeared about 1800, and only eight are still standing in the hearth, although several more once existed. Three are known outside the hearth area— two in Warren County, New Jersey, and one in New Castle County, Delaware. Most are medium to large structures and once seen, are not easily forgotten.

The Dornbusch-Heyl book cites a barn dated 1803 near Emmaus, and a barn off Cedar Crest Boulevard in South Whitehall in Lehigh County is dated 1806. The Furlong barn near Tylersport in Montgomery County is dated 1810. It seems that the barn style was not erected past 1840.

A very large stone-posted barn dated 1806 in the Jordan Valley in Lehigh County is one of the earliest barns of the type in the hearth.

The small, three-bay barn in West Rockhill Township has three stone arches but has been considerably altered on its interior.

The circa 1810 Furlong barn in Bucks County was built into a natural earth bank.

## Defining Characteristics

Stone arched barns combine characteristics of stone-to-peak Standard barns and English Lake District barns. They have distinct forebays and stone to the top of the front wall, as is seen on Lake District barns. The major difference is the stone arches at the front of the recessed stable wall. Other stone arched openings exist in other barns, but they are typically openings for wagon or cart sheds at the front of a barn. Stone arched barns have virtually all the other fundamental features seen in most other Standard barns. This includes roof structures, bents, mow-stead walls, and wagon floors, granaries, mow areas, and basement level features such as staked mangers and manger troughs, ceiling timbers including summer beams, joists, and stable walls including doors.

## Literature on Stone Arched Barns

No treatise has covered stone arched barns in depth. However, the Dornbusch-Heyl book includes a section on the barn style, Type K, where four of the barns are illustrated (Dornbusch-Heyl 1958). Three reside in Montgomery County while one (destroyed in 1954) was in Lehigh County. Heyl says that the stone arches, in conjunction with the stable walls, form "a kind of loggia & shelter" for the stable area.

Jeffrey Marshall devotes a two-page spread to this barn type, which he refers to as "stone arcade barns" (Marshall 2007). He includes a photo of a superbly built stone arched barn built in the mid-1830s. He also cites four barns in West Rockhill Township in Bucks County.

Arthur and Witney's book contains an illustration of the stone arched barn near Tylersport in Bucks County (Arthur and Witney 1972).

## Distribution

Since the early 1990s, I have identified eight stone arched forebay barns in Pennsylvania: three in Bucks County, two in Montgomery County, two in Lehigh County, and one in Northampton County. These few barns, plus two others in the Dornbusch-Heyl book, located in Montgomery County and Lehigh County, indicate almost conclusively that this barn style was constructed exclusively within fifteen miles of the Delaware River. For unknown reasons, this scantily represented barn-building tradition did not migrate west of Emmaus in Lehigh County.

It seems likely that a few others were constructed during the early nineteenth century. They have not been identified, or were dismantled or disintegrated. Three of the barns that appear in the Dornbusch-Heyl book are no longer standing.

## Dimensions

The range of dimensions of stone arched barns is somewhat narrow, perhaps because not many survive. Only eleven barns in Pennsylvania have been identified; ten had three bays and one barn north of Emmaus, destroyed in 1954, had four bays. I have measured eight of the barns. Lengths range from 36 feet to 75 feet. Just

The small stone-arched barn in Lower Saucon Township in Northampton County has three arches. It is a rare area barn of the type.

The Lower Saucon barn has a 5½-foot-wide forebay that is about normal for a barn of the type. In an odd area treatment, the stone on the one end wall extends to within about 6 feet of the roof peak.

two barns are longer than 70 feet. Four barns are in the 50–55-foot length. Average length of the eight barns is 54 feet and average width is 37 feet. Widths range from 30 feet to 45½ feet. Perhaps the expense of stone including labor costs limited their footprint, as none exceed 75 feet in length.

However, the barn one mile north of Emmaus must have exceeded that length. Heyl wrote, "This was undoubtedly one of the most impressive examples of this type of barn" (Heyl 1958). Amen to that! Its rear wall had a rare central panel of stone separating the doors of the paired threshing floors. The barn likely exceeded 80 feet in length.

The longer the barn, the more stone arches it has. Six barns have three arches while five barns have five arches. The arches themselves come in two basic shapes – rounded and elliptical. Eight of the barns have round-top arches, sometimes referred to as segmental arches—that is, they are segments of circles. The other three barns have elliptical arches or arches that are flattened at their tops. The wall thicknesses vary from 19–23 inches, the same as in other areas of the barns.

Widths of the forebays in five recorded barns vary from 4½ feet to 9½ feet. Average width is about 6 feet. This dimension is close to the average width of forebays on Sweitzers and other Standard barns (without front straw sheds).

## Outstanding Examples of Stone Arched Barns

Perhaps the best example of great stonework is the barn just off Route 611 on the Edison-Furlong Road in Bucks County. While not a big barn at only 55 feet long by about 35 feet wide, it has five stone arches. The first several feet of stone at each front corner may be the finest stonework in the entire hearth area. Another particularly good barn is the 1810 Furlong barn near Tylersport in Bucks County, with three stone arches.

While all the barns discussed in this chapter are real eye-catchers, most likely the stone-posted Standard barns are the most appealing to the public.

## Summary

Why would only a few select farmers create barns with such prominent front walls and stone arches? It seems likely that they were making a cultural statement by demonstrating their wealth. Certainly these walls provided great structural strength, but the fact that tens of thousands of forebay walls in Standard barns were frame proves that these great stone walls were unnecessary. However, no one is complaining.

# Chapter Nine.
# English Lake District Barns

The stone English Lake District barn off Route 401 in Upper Uwchlan, Chester County, is an excellent example of the barn type. The 1820-vintage barn has a typical pent roof over the stable doors on the front stone wall.

No barn style in southeast Pennsylvania has come close to the number of Swiss-German forebay barns built in the eighteenth and nineteenth centuries. However, another two-level banked structure was prominent in English-settled areas. This style is known as an English Lake District or ELD barn, a name credited to Robert Ensminger.

## Defining Characteristics

- The two-level banked structure has an upper floor level for storing farm crops and threshing grain, and a lower floor for stabling animals. The bank—either natural or man-made—appears at the rear where hay wagons entered through 11–12-foot-high double doors. The

The circa 1820 stone-to-eaves-level English Lake District barn in Worcester Township, Montgomery County, has an intact pent roof over the stable doors.

front eave wall is usually stone and contains openings. A few frame ELD barns, likely built in the post-1840 era, were also constructed. This framing occurs above the stone foundation walls.

- Above the stable doors on the lower level is a barn-length pent roof that projects about 4 feet. Original condition pent roofs are quite rare. The pent roofs protected the stable doors and farmers from the weather. A rare ELD barn, in Durham Township in Bucks County, has two pent roofs—one a regular eave wall pent and one end wall pent.

- These barns were never built with forebays at the front, at least not originally. However, their framing and masonry construction are similar to forebay barns.

## Literature

Shoemaker, Dornbusch-Heyl, Ball, and Long have made observations about ELD barns in Pennsylvania, though they did not use the term English Lake District.

Shoemaker uses the term "two-level stone barn without a forebay" for the three barns covered. Heyl devotes an entire chapter to eleven barns that he calls Type E, and calls it an "all stone masonry two-level barn." Nowhere does he suggest a connection with English building traditions.

Ball references Henry Glassie, who stated that some bank barns in Lancashire ". . . have, instead, pent roofs over the stable doors. There are barns like that in Chester County. They are not common . . ." Here is the first direct association with English Lake District barn building customs originating in Lancashire, just south of the Lake District.

Except for Glassie's statement, it was not until Ensminger began researching the Pennsylvania forebay

The stone-to-peak English Lake District barn of the early nineteenth century in Springfield Township, Bucks County, is very large at 80 feet long. It is also rare in that it has four bays while most ELD barns have three.

barn in the mid-1970s that any scholar seriously considered the place of origin of this rather strange two-level banked barn that had no forebay. Ensminger first learned of the barn style from the work of Dornbusch and Heyl, who believed that the ELD barn (they did not use this term) contributed directly to the development of the Pennsylvania barn. He also knew that Glassie thought certain structural elements of the Pennsylvania barn had a Northwest England origin.

Ensminger first visited the Lake District in England in June–July 1978. There he found "a strong resemblance to Type E barns especially in the southern section" (Ensminger 1992). He continues that they " . . . may be prototypes of those Type E barns found in the counties adjacent to Philadelphia which were largely peopled by English stock." He visited about fifteeen ELD barns and observed others from his car. Several barns had pent roofs. His subsequent visit in 1995 yielded more examples of this barn style, confirming his belief that these American examples originated in England. He concluded that the ELD barn had no direct bearing on the evolution and progression of forebay barn building traditions in Pennsylvania. However, almost nothing of scholarly note has been done on the ELD barn in the hearth area.

A barn in Springfield Township in Bucks County has several original pent arms. The interior timber stub extends a few inches to the back of the inner stone wall and appears to be wedged at its top and bottom sides.

In his book, *Vernacular Architecture of the Lake Countries*, Ronald Brunskill, a leading British architectural historian, states that the earliest bank barns of the ELD type were not built until the 1730–40 era in the Lake District in England. He adds, however, that the type probably did not predominate until many years later (Brunskill 1974).

In Joseph Glass's 1986 book, he describes his study that involved fifty-three survey grid points. He identified the ten closest bank barns of any type to each point, totaling 496 bank barns. Those without a forebay numbered twenty-five, barns without a bank and a forebay numbered thirty-four. Nearly 450 barns had forebays. Ten of those were in the southeast part of the state. Five were in English-settled sections near Philadelphia. We don't know if these were banked barns or Ground barns. Numbers of ELD barns compared to forebay barns were meager indeed.

The 2007 book *Barns of Bucks County* by Jeffrey L. Marshall and Willis M. Rivinus, has a four-page spread on the ELD barn that includes ten illustrations.

Hubert F. Jicha III wrote a ninety-nine-page treatise for his 1984 honor's thesis at the University of Delaware, "Bank Barns in Mill Creek Hundred, Delaware." Mill Creek Hundred is at the extreme northern end of Delaware adjoining Pennsylvania at the arc. He attempted to draw comparisons between the two-level bank barns in Mill Creek with barns in the northwest sector of England in Cumbria. Although Jicha admits that many differences exist between the barns in the two areas, a major one being the roof framing, several similarities are present. Among these are the following: the rectangular shape parallel to the ground, the one-bay threshing floor, multiple doors and windows on the basement level stable wall, and transverse stall orientations. No forebays are seen. He says that if any antecedents are found, they would be in Pennsylvania.

Jicha further states, "the search for direct antecedents may be an impossible task as the Mill Creek Hundred bank barn appears to be the result of the mingling of English styles based on a northwestern English form evolving into a uniquely American structure."

Tim Whitaker also did a study of bank barns in Cumbria published in 2001 in the *Journal of the Historic Farm Buildings Group*. He says that 749 bank barns were located in the area, which is by far the greatest concentration of the barn type in all of England. About fifty barns have pent roofs and all are in the southern section.

## Distribution

Although no systematic search has been done, the highest concentration of ELD barns is thought to be in the southern half of Bucks and Montgomery Counties and throughout Chester and Delaware Counties, all of which had the heaviest English settlement in southeast Pennsylvania.

I have identified fifty-three ELD barns: Bucks (twenty-four), Chester (nine), Montgomery (eight), Berks (three), Northampton (three), Lehigh (three), Delaware (two), and Philadelphia (one). No ELD barns have been reported in Lancaster, Lebanon, or Dauphin Counties, although a thorough search of the hearth would likely increase the count at least eightfold. Nearly all of Lehigh County has been surveyed, so the count there would not likely increase.

## Construction

Most of the ELD barns are stone. Similar to forebay barns, stone-to-peak ELD barns were likely the first built. ELD barns with stone-to-eaves-level on their end walls are probably not common. A rare construction type in the northern half of Bucks County includes stone on end walls to within about 5 feet of the peak, just like Standard barns in the same area.

Stone end wall barns are also fully stone on their front and rear walls in line with end bays or mows.

A rare frame barn, non-extant, occurred near Hosensack in southern Lehigh County. It was built circa 1860. It is not known how late frame ELD barns were constructed. Some of these might have been built in the last decade or two of the nineteenth century. Another barn along Route 82 in West Brandywine in Chester County has stone on its end walls but is strangely frame (likely original) on its front wall above the basement level.

One amazing tale about Quakers concerns an apparent ELD stone barn built in 1791 by Algeron Roberts in the Pencoyd area of Lower Merion Township in Montgomery County, not far from Philadelphia. He was "censured for such extravagance" by Friends Meeting congregants (archives of the Roberts family, Historical Society of Pennsylvania). Supposedly such stone barns were not normally constructed in that area and the story offers extraordinary insight into the dynamics of barn-building traditions of the late eighteenth century.

## Coincidence of Earliest Bank Barns—Forebay Barns and English Lake District Barns

ELD barns were built over a period of one hundred years—about 1770 to 1870. The oldest surviving ELD barn almost exactly matches the oldest known forebay barn (1770 stone Beiler barn in West Lampeter, Lancaster County) remaining in the state. It is true that bank barns of both styles may well have been built a decade or two before 1770, as changing agricultural economies in the third quarter of the eighteenth century dictated the building of two-level bank structures.

## Era of Construction

The earliest identified ELD barn is the original section of the Wickersham stone barn in southeast Newlin Township in Chester County. This remnant section is part of a much larger stone Double Decker barn built in 1809. It has two intact stone end walls; one stone has a carved date of 1769. Part of the front wall has prominent pent arms. Forty years later most of the structure was removed when it was expanded as a Double Decker in 1809.

The third oldest example is the 1775 Bartram Gardens stone barn on the west bank of the Schuylkill River in Philadelphia. This is a medium-sized structure whose authentic interior framing has been largely eliminated, though its original exterior stone walls remain.

These two barns, six years apart in age, are about twenty-five miles from each other—a fair distance in the era just before the Revolutionary War, suggesting that other barns also appeared in the region.

The fourth earliest dated barn is the remnant 1777 Wilkinson stone barn in eastern London Grove in Chester County. The barn was later expanded to a Double Decker. Only the roof structure was removed, and the upper floor was converted for sink mows and grain compartment bins. The wagon floor with its original planks remains.

A rare double log crib three-bay ELD barn was in Tredyffrin Township in Chester County. It was removed in late 2000 and dendro-dated at about 1780. Its dimensions were 46½ feet long by 30 feet wide. The upper floor's exterior logs were oak, up to 15 inches in diameter, and were left in the round.

The Wilkinson Double Decker barn, dated 1777, started as an English Lake District structure and is the third or fourth oldest known surviving barn of the type. The front cover barn was built in 1770 (the second oldest ELD barn).

The English barn at Bartram Gardens in West Philadelphia was built in 1775. It has very small vertical slits and the original interior framing is almost completely gone. It may be only one of three vernacular barns left in the city.

The Wickersham barn apparently began as an English Lake District barn in 1769, according to a date-stone on the one stone end wall. In 1809 the barn was greatly expanded as a Double Decker.

The brownstone Fatima barn in northern Bucks County has had some modifications on its front wall. The stone drip ledge across the full front marks the upper edge of the former pent roof.

After about 1790, ELD barns became increasingly common. Until about 1860 most were stone. Some built between 1770 and 1870 were likely frame. After about 1850, they were built less frequently.

Like many forebay stone barns, a number of ELD barns have vertical slits or splayed loop-holes. Probably due to similar economic conditions in various areas, ELD stone barns, like most stone Standard barns, likely were no longer being built after about 1875.

## Exterior Dimensions

Twenty-one ELD barns were measured for length and width. The length of one barn could not be recorded. Nineteen barns had three bays and only two had four bays.

Of the three-bay barns, the average length was 51½ feet while the average width was 33½ feet. Range of lengths is 39 feet to 70 feet. Eleven of these barns were longer than 50 feet. Seven were between 40 to 49 feet wide. One barn was less than 40 feet long. Widths ranged from 26 feet to 40½ feet. Three of the barns were more than 40 feet wide. Twelve were 30 to 39 feet wide. Three were less than 30 feet wide.

Of the four-bay barns, the average length was 62½ feet. The average width was 34½ feet. The longest barn, in northern Bucks County, was about 80 feet long and is the longest known ELD barn. The other four-bay barn is quite remarkable in that it is just 45 feet long. For a banked barn of any type, such a length is remarkably short and unusual.

On average, the largest ELD barns have a smaller footprint than the largest forebay barns, both Sweitzers and Standard barns. This assertion may be made given the unusually large barns that occur in much of Lancaster County and parts of Lebanon County and western Berks County. It seems that in many cases, German-based farmers settled the most fertile lands. These places of higher crop yields demanded the biggest barns.

Building costs may also have been a factor in the smaller ELD barns, as a one-hundred-foot-long ELD barn or longer would require an extensive amount of stone on the front walls.

These constraints may also have dictated the appearance of Double Decker barns. (See Double Decker chapter) Double Deckers with an extra floor level may be regarded as vertical extensions of ELD barns. The barn style minimized the footprint and likely saved money on labor and materials.

## Interior Framing and Roof Structures

Interior framing matches the framing in forebay barns, especially the Standard style. This includes both upper floor bent constructions and basement ceiling timber support structures. However, some of the basement

The Kline barn circa 1800 in Upper Milford, Lehigh County, has longitudinal basement ceiling joists in one area. Other English barns possess this feature. Some Sweitzers in Lancaster County also share this feature.

joist orientations are longitudinal, rather than transverse as in forebay barns.

Roof structures are basically the same as those in Pennsylvania forebay barns. The relative incidence of the three major rafter systems, varies, however. Early style German roofs have been identified in just three ELD barns and only two are extant. One is in a frame circa 1800 barn at Pennypacker Mills in Schwenksville, Montgomery County. The other is a 1770s stone ELD barn north of Moselem Springs in Richmond Township, Berks County. The third barn, removed in 1998, was on the Heller homestead in Hellertown, Northampton County. It seems a discordant note that German construction should be incorporated in an English barn.

Perhaps even stranger is the fact that only four ELD barns (out of fifty-four surveyed) have English-based Principal Rafter Systems. This brings to question many long-held ideas about cultural connections between different types of barn construction. Principal Rafter Systems have been incorporated in English buildings in Pennsylvania since at least 1700. The original section of the circa 1695 Merion Friends Meeting House in Lower Merion Township, Montgomery County, features this roof support type, as does the 1704 Harriton house in Bryn Mawr, Montgomery County.

## Granaries

Granaries appear in the same relative locations as in Sweitzer and Standard barns. They occupy the full widths of one end bay and are adjacent to the front walls of the barns. Ceilings are about 7 feet high and vary from 9 to 12 feet wide. Passageways are about 4 feet wide at the front side of the barn, and grain bins appear at the back. The incidence of original granaries, with their ceiling boards, rear wallboards, and grain bins are about the same as in Standard barns.

## Outstanding Examples

### Wickersham Stone Barn

The barn in southeast Newlin Township in Chester County, at the Wickersham homestead, is apparently the oldest example, actually a remnant, of an ELD barn anywhere. The remnant structure and its later counterpart, an 1809 built Double Decker, is recognized as one of the great barn structures in Pennsylvania. A photo is included in the classic book *The Barn: A Vanishing Landmark* (Arthur and Witney 1972).

The original section retains its two stone end walls and part of its front wall. A date of 1769 is seen on a stone several feet below the original peak on the one end wall. At least two substantial pent arms at the front survive. The original barn measured 64 feet long by about 33 feet wide. In 1809, the barn was converted or greatly expanded into a Double Decker that has a few remarkable features. (See Double Decker subchapter for more information on this barn.)

### Tredyffrin Double Log Crib Barn

A barn near Tredyffrin in Chester County was an extreme rarity. It was removed for reconstruction at a nearby site in 2000. The timbers were dendro-dated to 1779, thus the barn was likely built in 1780, making this a Revolutionary War–era barn. It had three bays with two double log cribs and measured 46½ feet long by 30 feet wide. Exterior logs were left in the round and a few approached 15 inches in diameter. It had a log stairwell (with log walls) that led to the basement. One wall retained an original oak door. It seems likely that other ELD double log crib barns existed, but this is the only one known to scholars.

An English barn in Richmond Township in Berks County has a date-stone of 1770. The interior structure includes a German-style roof and has rare plank mow-stead walls complete with marriage marks.

## Moselem Springs Barn

The 1770 barn in Richmond Township north of Moselem Springs, off Route 662 in Berks County, is another outstanding structure. It has three bays and measures 57½ feet long by 30½ feet wide. It has splayed loopholes on its end walls. This is one of the rare ELD barns with a *liegender Dachstuhl* roof support. The roof has two classic trusses where each truss retains its original timbers. The barn has an original longitudinal *oberdenn* beam over the wagon floor that has nine transverse "joists" at each side. This construction is unique for any type of barn.

Another exceptional aspect is the two mow-stead walls with seventeen horizontal planks, rather than boards. The planks are 2½ inches thick and up to 21½ inches wide. The ends of the planks are positioned into vertical grooves in the bent posts and have marriage marks matched to the marriage marks on the posts. Such planked mow-stead walls are rare. Curiously, the other rare ELD barn with a *liegender Dachstuhl* roof support is in Pennypacker Mills in Montgomery County, and also has plank mow-stead walls. The only other barn with such walls is a stone Sweitzer southwest of Dauberville in Berks County, and probably dates to the first decade of the nineteenth century.

## Kline Barn

The Kline barn near Shimerville in Upper Milford Township, southern Lehigh County, was built circa 1800 or earlier. It is one of two remaining ELD barns in the entire county. It has three bays and measures 63 feet long by 36½ feet wide. It has stone on its end walls only to the eaves level, which is peculiar for such an early barn, and splayed loopholes on its exterior walls. Its original pent roof is gone.

The roof structure is a Principal Rafter System with five principal rafter pairs. All timbers are oak. The mow-stead walls are composed of the earlier style oak

The stone-to-eaves-level Kline barn in Upper Milford in Lehigh County has lost it original pent roof. The barn has short vertical slits that function as interior ventilation.

boards secured with wrought nails. The one mow-stead wall door is also oak and has its original and intricate early style wrought iron metal latch. The one end bay or mow is a wide 29 feet, one of the widest ever seen. The basement at its east side has longitudinal ceiling joists that supplant the common transverse joists. The basement also includes a rare longitudinal stone wall that divides the east section into two rooms.

### Glick Homestead Hybrid Barn

A hybrid barn of unprecedented type is located at the Glick homestead in Upper Saucon Township, Lehigh County. A hybrid building combines traits of two or more ethnic building customs. The Glick barn is a three-bay structure that combines a two-bay Sweitzer section with a one-bay English Lake District (ELD) section. Dimensions (without forebay or pent roof each 6 feet wide) is 66½ feet by 30½ feet. Each barn section has stone to the eaves level at their end walls. The Sweitzer section includes the wagon bay and one end bay or mow while the ELD section consists of one end bay that likely included a large granary. In addition, the full length of the barn has a Principal Rafter System and also retains both its original oak mow-stead walls and its original wagon floor planks. It seems most probable that the barn (both sections) was constructed between 1800 and 1825. This hybrid construction is all the more startling, as this area does not harbor, with rare exceptions, either of these bank barn styles.

## Summary

The English Lake District barn served as a counterpart to the German forebay barn style for at least eighty years, although the forebay barn far outshone the English Lake District style. After the middle of the nineteenth century, frame forebay barns grew in popularity, and building stone ELD barns was no longer practical. Especially after the Civil War era, the forebay style was the builders' choice for two-level bank barns.

# Chapter Ten. Construction Elements

Almost all 1700s and 1800s barns in the Pennsylvania hearth share common construction elements. The only conspicuous exception is the log crib walls of log barns. Virtually all barns were built from wood timbers, stone, and metal, although little metal was used in log barns, especially before 1775. Some barns used brick, mostly after 1830.

Not all the construction elements discussed on the following pages are equally represented in the various barn styles and types in this book. Early log barns, for example, usually don't have mow-stead walls. And queen posts are not usually associated with German roof trusses but are present in barns with Common Rafter Systems.

## Common Rafter System

This rafter system, seen across all construction eras, refers to a roof support structure consisting of same-sized rafters arranged in pairs in any given barn. They are usually pegged at their top ends and over-sail the wall plates at their bottom ends by 8 to 10 inches. They are rarely less than 4 inches by 4 inches or greater than 7 inches by 7 inches.

The wall plates are double notched to receive similarly notched rafters. These notches prevent outward thrust. Many pre-1810 or 1820 vintage barns have slotted wall plates that receive the rafters' cogged bottom ends.

The early style or pre-1810 rafters in most cases are straight in their extensions beyond the plates. In post 1820 to 1830 barns rafters are "shaped" or curved beyond the plates. This straight or curved aspect of rafter bottoms is often a useful general diagnostic dating tool. Hewn rafters are either square or tapered.

Many pre-1850 or 1860 era barns have hewn rafters; barns built after 1850 or 1860 often have milled rafters, but there are exceptions. In the greater Philadelphia area, a number of early nineteenth century barns have milled rafters, perhaps because of the easy access to saw mills.

It seems most likely that the earliest barns in the hearth had Common Rafter Systems, as they were small and needed the least support.

## *Liegender Dachstuhl* Rafter System

This early German style roof support is similar to the Principal Rafter type. *Liegender Dachstuhl* (LD) is German for "lying chair." Fifty-six barns have been discovered with this unusual and distinctive roof type. Normally, this early roof style is not associated with nineteenth century barns. Some of the earliest German buildings, including the 1719 Christian Herr house in Lancaster County, have LD roof systems. They have origins in Europe that extend back to the late fourteenth century.

This German roof system was far more labor intensive than the Common Rafter System and Principal Rafter System, which is why the German system was largely outmoded after 1800. The Industrial Revolution was fast approaching and with it, a move toward structures that could be built quickly and efficiently.

LD roof systems appear in early barns from about 1750 to 1790 but they have also been seen in Standard barns as late as 1850. A basic breakdown of barns with LD roof supports in the hearth area is as follows: stone Ground barns (4), frame Ground barns (1), stone

A frame Standard barn in Greenwich Township, Berks County, has a Common Rafter System with original hewn rafters. Canted queen posts support the canted or angled purlin plates that support the rafters.

A stone Sweitzer in Longswamp has common rafters cogged at their bottom ends that are received into transverse slots. Slots are a diagnostic feature indicating a pre-1800–1810 era of construction.

Sweitzers (22), stone and frame Sweitzers (1), log and frame Sweitzers (1), frame Sweitzers (1), log Sweitzers (1), stone Standard barns (12), frame Standard (3), stone English Lake District barns (2), and frame English Lake District barns (1). This roof system was also found in a nondescript frame two-level bank barn and a frame Ground barn that had likely originally been a Sweitzer. The roof type has been identified in six of the hearth counties; it hasn't been seen in Bucks, Philadelphia, Delaware, or Dauphin counties.

It's interesting that at least three non-German-type barns had the LD roof. This is more evidence of the mixing of German and English styles.

Such LD roof trusses consist of a few structural members. The defining timber is the so-called "truncated Principal Rafter." These half-roof-length inclined timbers taper from 12 inches to about 6 inches wide at the bottom end. Their tops are prominently notched to receive barn-length purlin plates. These rafters are always associated with Common Rafters above them. A horizontal straining beam joins the Principals (in a pair) at their top ends. In addition, stout end braces join the Principals with the straining beams. Immediately above the straining beams are collar beams that join the Common Rafters in a pair. This is referred to as a classic truss. A variant type of truss lacks either end braces or a straining beam. Usually post-1800 trusses are the variant type.

Two trusses of a German roof structure appear in a circa-1780 stone and frame Sweitzer in Upper Milford, Lehigh County, that has most of its original timbers. It has canted purlin plates rather than the normal "set square" plates.

## Principal Rafter Systems

More than 245 barns, including more than 125 Standard barns, have been identified with this English-based roof system. Several Ground barns and more than sixty

A Principal Rafter System in the 1849 Wagner Standard barn near Hellertown, Northampton County, has all the original timbers. The Principal Rafters extend from the eave wall plates to the roof peak.

A variant roof structure, in a late-eighteenth-century stone Sweitzer near New Holland, Lancaster County, is somewhat of a cross between a German *liegender Dachstuhl* type roof and an English type. These roofs are not common.

Sweitzers have the roof type. This roof type has large rafters that span the entire roof slope in line with transverse bents below them. In addition, a number of barns have these rafters about mid-way along their end bays. They are much larger in cross section than the common rafters seen between the Principal Rafters. Except for Delaware and Philadelphia Counties, this roof type has been seen in all the counties in the hearth area and several others west of the hearth in barns mostly of pre-1830 or 1840 vintage. Exceptionally, in the greater Saucon Valley near the Delaware River and south into the northern area of Bucks County, barns with this roof type were built until 1850 and perhaps even 1860, though the number is disproportionate compared to other roof types.

## Hybrid Roofs

A few barns in Lancaster County have what might be called a nondescript roof type. They can only be described as a hybrid between the German and English roof styles and appear to be original. One example occurs in a circa 1790 stone Sweitzer a mile north of New Holland. The non-tapered Principals extend only to the purlin plate level.

Curiously, a few barns have been found with spliced rafters. This unusual condition occurs at its juncture with the purlin plate. One such barn occurs in an 1815 English Lake District barn just east of Doylestown in Bucks County.

## Purlin Plates and Queen Posts

These posts and plates act in concert to support common rafters. Purlin plates are longitudinal timbers that stretch from end wall to end wall about mid-way along each roof slope. In some small barns, 30 to 40 feet long, plates stretch the entire length of a barn as single pieces of timbers. But in most barns, plates are spliced in one to three spots called scarf joints.

A few barns in the hearth have purlin plates that extend beyond the end walls by a few inches and are clearly visible on the exterior. Several of the earliest buildings in the German-American culture area, such as the mid-eighteenth-century buildings at Ephrata Cloisters in Lancaster County, have extended purlin plates.

Interestingly, on the upper floors in certain barns in the hearth, some splicing details such as "keys" or wedges, which prevent the two spliced sections from separating, are duplicated exactly in scarf joints of the basement-level summer beams. We may say with almost complete

Vertical queen posts are present in an 1830s frame Standard barn in Montgomery County. Such oriented posts were commonly included in roof structures in barns until about 1830.

The Beiler log Sweitzer in Elizabethville in Dauphin County has very rare extended purlin plates clearly visible on the exterior. Only a few other barns possess such plates.

A good number of keyed scarf joints are seen in various wall and purlin plates and also summer beams in barns. This example is in the Henneberger stone Sweitzer in Franklin County.

Canted or angled queen posts often appear in post-1830 or 1840 barns. Vertical queen posts are seen in some barns of this vintage.

confidence that the same timber framer fashioned both of these joints.

Since purlin plates stretch significant distances (end wall to end wall), they would sag. The remedy for this was queen posts, either vertical or canted. Both types emanate (at their bottom ends) from the upper ties of the transverse bents or framing units. The posts appear 8 to 10 feet from the eave walls. However, the manner in which plates are supported at end walls differ principally in two ways, depending on the materials used to construct the barn. In stone-to-peak barns, the plates were either inserted directly into the stone walls or supported in the normal way—by queen posts. In both stone-to-eaves-level barns and frame barns, posts support the plates.

Vertical queen posts are the most common form of plate support in barns built prior to 1830 or 1840. Certainly vertical posts may be seen on occasion in barns built far into the nineteenth century. In some early 1800s barns, the posts are "clasped" around the plates. That is, at their top ends, the posts have wood extensions at the vertical sides of the plates that add stability. Such clasped posts are not common. Some of these upper post extensions in certain pre-1820 or perhaps pre-1830 barns have a distinctive gun-stocked aspect. That is, the extensions have a flared-out or angled characteristic that creates a greater load-bearing surface. These angled extensions often join to the common rafters above them to offer greater support.

After 1830 or 1840, certain barn construction techniques changed in response to evolving farm economies. One change was the re-orientation of the posts from a vertical position to an angled or canted position.

The angled placement of the posts necessitated the angling of the purlin plates for a more secure fit. To be sure, canted posts are seen in pre-1830 barns, but this feature is not common. An early example is seen in the embedded queen posts in the stone end walls of the altered 1777 Wilkinson barn in Chester County. Some barns built circa 1800 have these canted posts. Builders started to angle posts after 1830 or 1840 to create more open space in the upper loft to allow easier loading and unloading of farm crops.

The orientation of queen posts helps determine general dates of construction. When I come upon a new barn and have just a few short moments to record things, I always note the disposition of the queen posts.

## Transverse Framing Units, or Bents

Transverse framing units, also called bents, on the upper floors of bank barns, extend from the front wall to the rear wall in almost all barns except log barns. A number of log barns have mows that include frame units adjacent to the wagon bays. Bents constitute the fundamental upper floor skeleton. Sweitzers also have bents, even stone end wall Sweitzers, but these bents do not extend into the front forebay. The fact that Standard barns have bents across their full width is a marked improvement, labor-wise, over Sweitzers. Building separate forebays in Sweitzers was labor intensive. The fact that the distinctive appendages were not part of Standard barns, while they still maintained the cantilevered effect in front, demonstrates the streamlining action as a cultural constraint that builders were then subjected to in the post-1820 construction era. Builders had to produce barns in the most cost-efficient way possible. Eliminating the front appendages of Sweitzers was a notable part of that process.

Dropped tie beams are most often found in post-1870–1880 barns. Ties were placed about a foot below the tops of wall posts.

The front 4 or 5 feet of bents in most Standard barns are contained within the forebay. Ensminger says that this cantilevered area is considered the front forebay of a Standard barn.

Gable ramp structures have bents set longitudinally. This bent orientation creates the aisles of these barns. The number of bents in a given barn varies according to the number of bays and the type of construction. For example, three-bay stone barns have two bents, while four-bay stone barns have three bents, and so forth. However, three-bay frame barns have four bents, while four-bay frame barns have five bents. The extra bents in frame barns include the bents that constitute the end walls. These frame walls, of course, do not exist in stone barns. Frame barns of equal bay numbers as stone-ended barns always have two more bents than stone barns.

Many bents consist of two eave wall posts (one at each end) and one to three posts along the lengths of the bents. One or two of the inner posts often form the vertical sides of built-in ladders. At the tops of the bents are upper ties or single full-barn-width timbers. Ties are not actually a part of many bents, as they were added to bents after they were positioned vertically on the upper floor or deck. When such upper ties are present, the disposition and lifting of the bents is referred to as "normal assembly." Many bents with three posts have upper half-barn-width girts a few feet below the upper ties and also lower half-barn-width girts 5 to 6 feet above the floor level. Often, braces placed at strategic places are also a part of bents. Some bents have double ties at the top separated by a foot or so.

The John Weiler homestead Sweitzer in Longswamp in Berks County has superb joinery, such as the ties of the bents joined to stout posts. The earlier-style angled notches or shoulders were used with marriage marks.

A close-up of the joinery in the Weiler Sweitzer in Longswamp indicates oak construction and square wooden pegs (trunnels) that secure the tie to post junctures in the bents.

Although the precise timbering of bents can certainly vary among Standard barns, bents in these barns offer far more uniformity than in other barn types, such as Sweitzers. On average, especially in earlier examples, bents in Sweitzers have fewer timbers than in bents in Standard barns. For example, in a rare three-bay stone barn in Chester County south of Downingtown, the two bents that flank the wagon bay have just one post each and also have, atypically, no built-in ladders. Each bent has front posts adjacent to the forebay, but the ties join directly into the rear wall stonewalls.

Many barns built after 1870 have "dropped" upper tie beams. These integral beams stretch the length of the bents. When the bents are placed into position on the upper floor, the ties are in place. In these cases, the process is referred to as "reverse assembly." They are not added to the already positioned bents like upper ties of bents, in cases of normal assembly. Upper ties in reverse assembly bents are 12 to 20 inches below the tops of the end posts.

## Two Means of Shouldering

In the vast majority of non-log barns built prior to about 1910, the joinery of timbers (beams) in transverse bents is accomplished with mortises and tenons. Ends of timbers have thin tongues of wood called tenons. They fit into pockets (voided wood areas) or mortises of the timbers that they connect to. Long pegs or wooden pins, also called trunnels or tree-nails, secure the joints.

Tenons fit into mortises in three principal ways, two of which involve shouldering. Shouldering refers to the point where the post or stud joins the girt or upper tie (beam). One type of shouldering is called diminished haunched shoulders or *niefalz* in the Pennsylvania German dialect. This term comes from Charles Speicher, an old-time timber framer of Womelsdorf, Pennsylvania. The end of the beam next to the tenon is angled and is received into a similarly angled cut in the post adjacent to the mortise. The bottom outer edge of the beam (just to the side of the tenon) rests on the "shoulder" or the bottom edge of the post to the side of the mortise. This takes considerable weight off the tenon and thereby increases the strength of the mortise and tenon joint. At the top, no corners are present. The *niefalz* is the earlier style and is seen in many barns.

The second type of joint is square shouldering because both the top and bottom corners of the beam-to-post union are square. The corners are connected, so to speak, with a straight vertical line. The ends of the tie or girt and the post or stud are also vertically cut.

In the third type of mortise and tenon joinery, the tenon of one beam fits directly into the mortise of the receiving beam. In this type of timber union, the tenon takes the full brunt of the weight of the timber union.

Often lower girts in bents join to posts with no shouldering, such as in the Kalmbach Park Standard barn in Macungie Borough in Lehigh County. Such joints had far less weight to carry than joints that required shoulders.

An interesting variation exists in the two types of shouldering in some barns, especially in Standard barns, which have both upper partial barn-width girts and lower partial barn-width girts. The upper girts often exhibit diminished haunched shouldering while the lower girts include no use of shouldering. There is a good reason for this disparity. The upper girts support the considerable weight of *oberdenn* beams that in turn support the great weight of farm crops above the beams. In the case of lower girts, no weight is basically carried. In one sense these girts are "free floating." The lower girts accomplish two things: they increase transverse stability to the bents, and the lower girts provide nailing surfaces of the top ends of the mow-stead walls.

## Handmade Barns versus Mass-Produced Barns

Barn-building specialists adhered to certain guidelines in the 1700 to 1900 era in North America. Suffice it to say that framers fashioned timber joints in particular ways in different eras. Certainly, methods of different time frames overlapped to a considerable degree.

Before about 1810, in many or most areas in the east, timber unions were joint specific. For example, where certain ties in bents joined wall posts, a particular tie was cut or formed in such a way that it could only be united to a particular wall post. In other words, the tie could not be joined to any other post. This process was called the scribe rule method of timber framing. Marriage or carpenter marks were usually used (applied to the timber unions). This was a labor-intensive and painstaking manner of making barns. These were truly handmade barns.

But then a changeover occurred after 1810 or 1820 where builders entertained the concept of interchangeability of corresponding parts of timber frames. This later method might have started in the last decade of the eighteenth century. This was likely an age of experimentation where certain timber connections were fabricated in certain ways. This procedure was considerably less labor-intensive than the scribe rule method. Marriage marks were either not used in this process or sparingly so. This method was a distinct improvement over the scribe rule method, at least as far as speed of execution was concerned. To quite a degree, barns from this era were essentially "mass produced."

One probable impetus of the changeover of timber framing practices from early scribe rule era barns to later square rule era barns was the gradual depletion of large, old-growth timbers. The abundance of large timbers allowed for freedom of joinery in that framers could waste timber to a degree. But when the Industrial Revolution was in full swing after 1820 or 1830, they had to use smaller timbers, and timber frames and bents became standardized or more uniform. Definite exceptions to this existed when smaller timbers were probably used prior to 1800 or 1820 at certain barn building sites and larger timbers were used after 1800 or 1820 at other barn building sites. It is a certainty, however, that many barns, particularly in the hearth area and beyond in Pennsylvania, built well into the second half, and even the last third of the nineteenth century, were fabricated using the scribe rule. Marriage marks are often present in these barns.

## Built-in Ladders

Bents adjacent to wagon bays usually have built-in ladders that ascend to the upper ties. The ladders are often centered on the bents. Sometimes the ladders are off-center by several feet. In a fair number of cases, built-in ladders ascend to the roof peak. Built-in ladders

Built-in ladders of bents in many barns have lower sections that have cut-outs in mow-stead wall boards. The upper sections consist of vertical sides (posts or studs) and a series of rungs, most often of oak.

A minority of barns have built-in ladders that have good-sized side posts. Some barns have ladders with side studs.

basically fall into two major categories. The first type has rungs inserted in the posts, two at either side, or a post at one side and a smaller vertical stud at the other, from near the wagon floor up to the bottom of the upper tie. The second type has ladder rungs only above the mow-stead wall. This second type has 3-inch or 4-inch rectangular or semi-circular cutouts into which the farmers placed their feet. In a number of cases, these cutouts were one on top of the other on mow-stead walls. Possibly in a majority of cases the cutouts alternate side to side—one foot at each side as the farmer ascended the ladder.

Some ladder rungs were left in the round but squared off and set on the diamond as the rungs entered the post or stud. In other cases, the entire lengths of the rungs were square but "set on the diamond." This orientation rendered greater strength to the rungs than if they were left square. Diamond-shaped rungs are also more aesthetically appealing than square rungs.

## Mow-stead Walls

Mow-stead walls flank wagon bays. These walls consist of horizontal boards, usually oak or pine and sometimes tulipwood or walnut. The boards are 8 to 15 inches wide. A few barns have mow-stead wallboards more than 20 inches wide. Most boards in pre-1860 barns have non-beaded edges. But after 1860 or 1870, there were more mow-stead wallboards with beaded edges. This is a general diagnostic dating tool. In addition, barns built after about 1875 have rather narrow boards,

The 1835 Marcks stone Standard barn in Lower Macungie, Lehigh County, has two original mow-stead walls with the often-seen horizontal pine boards. At the wall bottoms are wide baseboards.

Nearly all mow-stead walls in barns have doors that lead to granaries, basement staircases, or access points between wagon bays in barns with four or more bays. Original doors on these walls are not rare.

The Hiestand stone Standard barn in Upper Milford in Lehigh County retains its original mow-stead wall with built-in ladder. The transverse bent here illustrates the use of posts, partial-barn-width girts and braces.

4 to 5 inches wide. In two extraordinary cases—near Doylestown in Bucks County, and in Lower Macungie Township in Lehigh County—the barns have vertically oriented mow-stead wallboards.

Mow-stead walls are 5 to 6 feet high and form a barrier to protect stored crops in the end bays or mows from activity on the wagon floor. Part of this is also to prevent the threshed grain from entering the mows. The walls also include doors that lead to a basement staircase or granary, or provide access from one wagon bay to an adjacent wagon bay in four-bay barns.

## Mow-stead Wall Nails as Dating Tools

Four major nail types are seen in mow-stead walls, and are an excellent general dating tool. The first and earliest style is the wrought or rose-headed nail. These nails were fully handmade and appear in barns from the mid-eighteenth century (earliest barns now extant) to roughly 1810. A few dozen barns have been identified with these nails. A second style is the transitional nail used from about 1800 until the second quarter of the nineteenth century. These nails have heads that were often made by hand. Several dozen of these barns have also been recorded. The third style is the cut nail type. This type with rectangular shanks is by far the most prevalent. Many hundreds (or even thousands) of barns retain this nail type in their mow-stead wallboards. Machines usually made this nail type in the 1820–1900 era. The nail heads are square. The fourth nail type is round-headed and was made from about 1890 to the 1920s. The posts and studs of bents were rabbeted on their edges where the boards were placed onto these recessed edges.

## Exceptional Mow-stead Walls

Occasionally, mow-stead walls have an extra door or two, especially those adjacent to end bays. In these cases, centered doors lead into full end-bay-width longitudinal passages with sheathed walls at either side. They measure about 8 feet wide and may have been used for the temporary storage of bags of grain. At the outer end of the passages are often gable end wall windows.

One exceptional mow-stead wall appears in a unique stone-to-peak Standard barn dated 1834 in Lower Saucon Township, Northampton County. Its upper floor structure is an amazing visual experience. One mow-stead wall has six original doors. Rear to front, the first door leads to the basement. The next door leads to a rare rear area granary. The third and fourth doors are side-by-side and lead to a longitudinal end bay passage. The last two half-height (3 feet high) doors are also side by side. The one door leads to an angled wood shoot through which farm crops were dropped in the basement.

The last element seen in mow-stead walls is the addition of baseboards at the bottoms of the walls. These boards are about 8 to 10 inches wide and are secured with one of the four nail types.

## *Oberdenn* Beams

The German dialect word *oberdenn* is derived from two words: *ober* (over) and *denn* (threshing floor). Alternate dialect spellings are *owwerden* and *iwwerden*. *Oberdenn* beams are a series of longitudinal timbers 11 to 12 feet above the one wagon floor in a three-bay barn or above one and sometimes two of the two wagon floors in a four-bay barn. These beams often rest on the partial-barn-width upper girts of the bents that flank the wagon bays, especially in Standard barns. In a normal 35-to 40-foot-wide Standard barn, a full complement of *oberdenn* beams number eight to ten timbers averaging 8 to 12 inches wide. Many times, these beams were laid loosely on the girts, and their weight held them in place; other times they were pegged into place. Many barns retain a few of these beams, though less than ten percent of barns have all of them.

In four-bay barns, *oberdenn* beams weren't always positioned in each wagon bay. Only one wagon bay was normally used for threshing and the other wagon bay was apparently used to store crops from floor to ceiling. Therefore, no *oberdenn* beams would be in place. In fact, a fair number of four-bay barns have a wagon bay with a bent at one side that has no partial-barn-width upper girts. This signifies that no *oberdenn* beams were in place and that crops were indeed stored from floor to roof.

Wide boards were transversely placed above *oberdenn* beams for storing crops. It is rather rare for a barn to retain even a few original *oberdenn* boards, as they were likely removed and used for other purposes. The best example of a barn with most of its original *oberdenn* boards in place is a two-bay Standard barn in Lower Saucon in Northampton County. They appear to be chestnut and range from 26 to 30 inches wide.

The frame Standard barn in Ruscombmanor in Berks County has a nearly all original *oberdenn* in its wagon bay. Boards, normally placed above the *oberdenn* beams, may be original in this barn, at least in part.

## Bays

Sweitzers and Standard barns have a variable number of bays or areas on their upper floor levels. Bays are the areas between adjacent timber frames or between frames and end walls in stone or brick-ended barns. Barns can have two to five bays, and in rare cases, six or seven. Most Sweitzers and Standard barns are three-bay or four-bay structures. All the barns have wagon or drive-in bays and end bays or mows. Three-bay barns have one wagon bay and two end bays or mows. Four-bay barns have two wagon bays and two end bays. A wagon bay was the area where threshing was done, the act of separating seed from the chaff. In barns with four or more bays, only one wagon bay, according to a few farmers in the hearth area, was used for threshing. In log barns, each log crib is considered one bay.

## Granaries

Granaries are usually at the front of the barns in an end bay. In Sweitzers, the granaries occupy the width (rear wall to front wall) of the forebay. In Standard barns, they occupy the entire width of the forebay plus an extra 4 to 6 feet to the rear of the forebay area. Granaries extend the full width of an end bay, from the adjacent wagon bay to the end wall. Granaries normally have 7-foot ceilings and a front passageway about 4 feet wide plus grain compartment bins—three to five depending on the barn's size. Some Lehigh River Valley granaries have four or five grain bins at either side of a central passageway. This condition offers tremendous storage capacity and often reflects a farm of extensive tillable acres. A fair number of large barns, more than 70 feet, have two granaries, one in each end bay.

Granary entry doors, many of which are original in various barns, are in line with and part of mow-stead walls near the front of the barn. It is quite common to

Tally marks set in bundles of five are often seen in or near granaries. The unusual stone bank barn in Longswamp, Berks County, is rare in that the tops and bottoms of the bundles have long, continuous lines.

see rows of tally marks scribbled on the walls or entry door counting the bushels of grains handled.

Granaries often retain their original ceiling and rear wallboards. Some barns have granaries with doors at their front or forebay walls.

In certain early Sweitzers, granaries may not have been part of the original fabric. In these cases, granaries were often located in garrets of the homestead houses. One excellent example can be seen in the 1769 Handwerk stone house near Germansville in northern Lehigh County.

## Crop Storage

Crop storage sections in nearly all vernacular bank barns were in the end bays, floor to roof or near the roof, and 10 to 12 feet above the wagon bay in three-bay barns, in the space known as the *oberdenn* that was devoted to crop storage. In addition, in barns with four or more bays, one or more of the non-threshing wagon bays was also apparently filled with crops floor to roof.

In one-level Ground barns, crops were also stored in end mows above the stable areas. These barns also have *oberdenn*s where crops were stored.

The hay hole contrivance in a stone Double Decker Standard barn in Chester County is virtually all original. Sides are each about 3½ feet. Mortise and tenon joinery unite the verticals with the horizontal wood pieces.

## End Bays

Most barns, both one-level and two-level, have end bays or mows of unequal widths. However, some may vary by 2 feet to 7 feet. Why do such disparities exist? Some hearth area farmers have suggested that some mows were bigger to store larger amounts of a certain kind of crop. These width differences can easily be observed on barn exteriors and exist in the early 1750s Taylor English Ground barn in Chester County.

## Hay Holes or Funnels

Hay holes are often present in two-level bank barns. These are four-sided structures usually located in one of the end bays near a wagon bay and adjacent to the rear wall. The structure generally consists of three poles and a post of the eave wall in a frame barn. The post acts as a fourth pole. The structures measure about 3½ feet on each side. Occasionally hay holes were placed next to a mow-stead wall several feet from the rear wall. They acted as funnels where hay or other farm crop was dropped to the basement. Some of these hay holes appear to be original and others may have replaced the originals.

## Floor Planks

Wagon floors consist of wooden planks that nearly always span the entire width of wagon bays. In certain barns, the planks are tapered. When this occurs, the planks are staggered, or arranged with alternating wide ends adjacent to narrow ends at each side of the bays. In a rare treatment, one circa 1810 Sweitzer in Longswamp Township, Berks County, has a floor with a median sill that receives half-wagon-floor-width planks on each side.

Plank widths vary from 6–25 inches and are usually 2 inches thick to withstand the heavy loads and stand up to the wear and tear of wagon wheels. The species of choice seems to have been oak and hemlock, the latter especially in post-1850 barns. Planks are secured with wooden pegs, long nails, or metal spikes to the basement ceiling joists. Often, adjacent planks are joined to each other on their long sides via long, thin wooden splines. The splines were inserted into grooves on the edges of the planks and near their top surfaces. A few differing thoughts exist on the function of splines. One idea is that the splines offered the planks stability so they would not separate and create an uneven floor. Another thought is that they prevented threshed grains from dropping into the basement.

Many original wagon floors survive in two-level bank barns. However, in most cases, the rear 3 to 5 feet were subject to tremendous wear and tear from entering and exiting hay wagons. In addition, these rear areas are far more subject to the effects of weather including rain and snow. The planks here often rotted and needed to be replaced. The 1849 Wagner barn near Hellertown is a rare example where the last few feet of wagon floor are sound.

Original floorboards in end bays often survive and consist of 1-inch-thick boards up to 18 inches wide. Like the planks in wagon bays, they stretch side to side.

Viewed from the underside, in the basement, replacement planks and/or boards are usually easy to differentiate from original wood pieces.

## Basements

Farm animals were stabled in the basements of bank barns. At the far end of the barns (away from the house) where cows were stabled, two or three barred windows often punctuate the walls. On the near end walls, one or two split doors are often positioned near each other at the rear half of the wall. When one door is present, the door leads to a feeding alley. When two doors are present,

The wagon floor in the Weiler Sweitzer is all original, including half-wagon-bay-width planks that join to a rare median floor sill. Just one other known barn, in Franklin County, has such a floor arrangement.

The Bertolet stone Sweitzer in the Oley Valley, Berks County, has a nearly original wagon floor. Several of the generally wide planks are an unusual 22 inches wide and appear to be tulipwood.

The Wagner Standard barn near Hellertown has two end wall basement-level doors with original overhead hoods. The decorative curves at the bottoms of the hoods are quite unusual.

the one closer to the rear corner is a stable door and the other is a human door that leads to a feeding alley. One or two barred windows are also seen on the wall.

Roughly seventy percent of Sweitzers lack end wall basement entries. In the hearth, at least eighty percent or more of Standard barns have these door entries.

On the interior surfaces of the house side end wall (where horses were stabled) are usually one or two several-foot-long beams placed into the stonewalls. In the beams are a series of holes for stout pegs where harnesses were hung. In addition, wooden curry boxes are often seen near the stable wall. The boxes, about 4 feet above the floor, are about 1 foot wide by 9 inches deep by 11 inches high. Some are fully open at the front and others are fronted by a board on the bottom half. Others are double—one on top of the other or side by side. These boxes contained items used to care for horses.

It is important to note that the three elements—*spriggel* bars in door entries, curry boxes, and in-set beams with pegs—indicate the horse side of the basement. Cow ends lack all three of these elements.

In the greater Lehigh and Saucon Valleys, much of the northern half of Bucks and Montgomery Counties, and the eastern half of Berks County, horses were almost always stabled at the end of the barn nearer the house. This gave farmers easy access from the house and lane. Also, farmers may have been more comfortable having the considerably more valuable horses close to the house. In some cases, prominent metal bars on the stable doors were locked in place to secure the horses. Horse thieves were common, even 200 years ago, and farmers took the necessary precautions.

Staked mangers and manger troughs where horses fed were often placed longitudinally (parallel to the roofline) in basements. The mangers and troughs in the cow areas were placed transversely at the far end of the barn. In the hearth area, original staked mangers and troughs are not common. One amazing exception

The DeHart stone Standard barn in Berks County was built in the 1860s and has an original curry box in the basement wall just to the side of the stable wall. These boxes stored items incidental to the care of horses.

to this, although not in the hearth, is the circa 1805 McCoy stone Sweitzer in Franklin County. An exception in the hearth is the Groh brick Standard barn dated 1859 in Jackson Township, Lebanon County.

Stone rear walls typically have straight vertical surfaces. However, in a number of cases, the walls in line with wagon bays are slanted. Their bottoms project beyond the wall by 12 to 15 inches. This accommodation offered considerably greater strength than straight vertical walls. When loaded hay wagons entered barns in wagon bays, a great deal of force was exacted upon the stone foundation in line with these bays. Continual action of the hay wagons on the walls would weaken them. A counteractive force was for the builder to angle the walls, thus reducing the effect of the forward forces caused by the loaded wagons.

## Summer Beams, Joists, and Posts

One or two summer beams typically stretched longitudinally across the full length of the barn. These timbers were about 10 by 10 inches in cross-section or greater, and supported all the ceiling joists or transverse timbers above them. Joists run from the rear wall to the front wall. Joists in turn supported the floor above. Several posts support the summers. From the dirt floor level to the soffits or undersurfaces of the joists is usually 7½ to 8 feet. This was the clearance needed for movement of both humans and farm stock and also to accommodate the staked mangers whose tops were just below the joists. Joists and summer beams were usually oak, and occasionally, tulip poplar.

Surviving staked mangers in barns in the hearth area are not common. Farmers placed provender into the mangers where farm stock fed. Cow mangers were placed transversely and horse mangers longitudinally.

The large circa 1800 stone Sweitzer in the Oley Valley has a rather rare intact metal bar across the one stable wall door. The function was to prevent intruders from stealing horses.

Joists were not always placed transversely in basements. Joists in most gable ramp Standard barns were longitudinally positioned below the middle aisles. This is also true of several Sweitzers in Lancaster County, such as the Treadwell barn near Strasburg, and some English Lake District barns have longitudinal set joists.

In a stone-posted barn in Chester County, a basement summer beam supports the joists above it, which in turn support the upper loft floor.

## Spriggels

Horse stable doors often had *spriggels* or horizontal wood bars drawn across the middle of the doors at night. Most of the *spriggels* are oak. The ends of the *spriggels* often had metal rings attached. *Spriggels* would be housed in wood boxes set in the stone stable walls adjacent to one of the door jambs. The *spriggel* was withdrawn when the metal ring was grasped and the bar slid out of the box and drawn across the door opening. The outer ends of the bars were placed in square holes in the door jamb opposite the *spriggel* box.

*Spriggel* bars were used in Ground barns, Sweitzers, and Standard barns. The use of *spriggel* bars was gradually reduced through the decades of the nineteenth century until about 1870 or so when they were rarely used.

The bars were 3-inch squares with boxes just big enough to accommodate them. Several barns in the hearth have double *spriggels*, especially in Chester County. A Standard 1850 barn in Upper Milford Township, Lehigh County, has two stable wall door entries, each with double bars. Some end wall stable doors also have *spriggels*. Another rare bank barn of an unusual type at the Kaufman homestead in Oley Township in eastern Berks County has three stable doors where all three have *spriggels*. It is not known when *spriggels* first came on the scene. The earliest barn with an original *spriggel* is the 1787 Bertolet stone Sweitzer in Oley Township, Berks County.

## Fudergongs

One to three transverse feeding alleys, or *fudergongs*, 4 to 5 feet wide were common next to stable areas. Farmers entered the barn to distribute food to the animals. When two doors were placed at the basement levels at end walls, one of the doors led to a feeding alley.

*Spriggel* bars were placed across door entries at night to contain horses. They were often on the house side of the barn. At the Weiler homestead in Berks County, a bar was placed on the opposite side.

A transverse feeding alley in a circa 1870 frame Standard barn in Lower Saucon Township, Northampton County, is almost fully original. They are usually about 4 feet wide or wider.

## Basement Floors

Floors of Standard barns were usually bare earth. Perhaps the ground was hardened by some means. Only occasionally were the floors stone or cobbled with small round stones. For example, a frame Standard barn built circa 1875 in Lower Saucon Township in Northampton County has a rare cobblestone-like floor in part of the horse stable. The Wagner stone Standard barn in Lower Saucon has a remnant wood floor in one of the horse stables.

The circa 1805 McCoy Sweitzer, in Franklin County, has a remnant wood floor in one of its feeding alleys. Dirt floors are usually flat. A few Standard barns in the hearth area have basement floors elevated by about a foot at the rear.

A very rare wood plank floor in the Wagner barn near Hellertown was used for positioning horses. This floor is not likely original to the barn.

## Wagon Doors

The very big paired doors (one to a side) on the rear walls of both ground barns, two-level bank barns, and Double Deckers are the entries that allowed loaded hay (farm crops) wagons into the various barns in the Pennsylvania hearth. Depending on the barn one to three wagon doors appear on rear walls, but most barns have either one or two wagon doors determined by the barn size. The most wagon doors seen thus far is five in a seven-bay barn in Lebanon County.

Perhaps one-fifth to one-quarter of all vernacular or pre-1900 built barns retain at least some part—to varying degrees—of their original wagon doors. A number of these doors have replacement wood pieces at their bottoms, as these areas are vulnerable to rot. Many dozens of wagon doors (on barns that survive) are nearly fully original. In certain instances, one wagon door half is original and the other original door half is gone, replaced by a modern door.

Some variability exists in the dimensions of wagon doors. Ground barns—generally decidedly smaller than bank barns—have smaller doors. For example, the small stone (ca. 1830) ground barn in Lower Macungie, in Lehigh County, has doors 11 feet high and 11$^1/_2$ feet wide (both door halves). Some of the biggest wagon doors on two-level barns approach 13 feet in height and 15 or more feet in width (both halves). All the doors had to be of sufficient height for loaded hay wagons to enter the rears of the barns. Original vertical boards on door fronts can be up to 15 inches wide. The boards are most often of pine.

Three basic types of wagon doors were installed in barns and their fine details vary to a great degree. The first and oldest type that go back centuries—used in European buildings and likely elsewhere—is the harr hung type, where the doors pivot on fair-sized poles or beams at the attachment sides of the doors. Pins at the ends—both top and bottom—are often received into holes in overhead lintels and sills, or stones at the bottom ends. (Other means of attachments were employed.) Only six or seven barns with harr hung wagon doors have been identified in all of Pennsylvania, at least three of which are seen on hearth barns.

The second type is the frame wagon door. In most cases long exterior metal hinges are not seen. Rather, the internal structure of the door frame is a fair-sized vertical wood piece or beam mortised in several areas to receive tenoned horizontal and tapered wood pieces, and most often diagonal brace-like pieces. All of the wood piece unions are secured with wood pegs (frame construction). Horizontal pieces are used instead of metal hinges to make the door rigid. Literally many hundreds of original frame doors have survived, or remnants of them have. Such frame doors were built at least from the last decade or two of the 1700s to as late as about 1870 (see top pg 47).

The third type is a board and batten, where long metal hinges are seen on the door exteriors. The metal hinges are most often secured to the inner horizontal battens by large screws, or are riveted. Diagonal battens are also present, and some of the larger doors have double parallel battens per each door side on each upper half and lower door half.

Many original wagon doors also have original human doors in one of the door sides (halves) that nearly always swing outward, as do all the main wagon doors.

Eave wall board and batten type wagon doors with exterior hinges are original to the Newmoyer stone to peak Ground barn built circa 1830 in Lower Macungie, Lehigh County. Barn retains its original human door in one wagon door half.

# Chapter Eleven:
# Regional Influences

The stone Standard barn in the Pektor Hill area of Northampton County has circular loopholes. These loopholes are confined to barns in the greater Saucon Valley about eight to ten miles west of the Delaware River.

A regionalism is a particular construction characteristic present to varying degrees in certain barns extending over variable geographic areas. Regionalisms exist in barns all over the hearth in ways big and small. A farmer would see a Principal Rafter System in his neighbor's barn and then use it on his own barn. A builder might tell the farmer about a success story at another farm. And sometime thereafter, the feature was repeated in a new barn, and then another. This is why similar features might appear in just a few barns or in many barns across variously sized geographic areas.

A log Ground barn circa 1790 in Tinicum Township in Bucks County demonstrates the regional aspect of inverted V-notching in log structures in the eastern half of the hearth area.

A log barn demonstrates the regional aspect of dovetail corner notching in log structures in the western half of the hearth area.

This process explains, for example, how Up-country Posted barns came to extensively dot the countryside north of Chester County. The idea of adding front straw sheds to existing Standard barns was widely successful. Other ideas never caught on substantially in certain areas. For example, the use of Principal Rafter Systems in the greater Saucon Valley occurred extensively over a distance of ten or twelve miles and then this concentration stopped, though it was used to lesser degrees in many other barns throughout much of the hearth.

Another region-specific feature was the circular loop-holes in the end walls and rear walls of stone barns within a few miles of Bethlehem. Similarly, brick patterns (usually sheaves of wheat) appear on a number of stone barns in the greater Lehigh Valley, extending into northern Northampton County and south into Berks County along Route 100.

The stone arched Standard barn was prominent within a twelve- to fifteen-mile corridor along the Delaware River. This construction feature, consisting of stone arches on the front walls, does not extend beyond this area in the hearth. The answer why, as so many things to do with barns, is lost to the ages.

Regionalisms even apply to the great Sweitzer barns that formed extensive colonies in the greater Lancaster area and nearby Lebanon County and western Berks County. East of this area, this building tradition lost momentum. And on log barns, dovetail notching was prevalent in the western hearth, while inverted V-notching was common in the eastern regions.

Travel constraints affected the characteristics of barns in different periods. The use of heavy stone, for example, probably depended on particular modes of transportation. If local sandstone, say in the area south of East Greenville in northern Montgomery County could only be conveniently transported a few miles, a regionalism of sandstone barns would have developed in a restricted area. But builders could incorporate particular joinery methods over much wider geographic areas, as they could manipulate timbers in any manner they wished.

Regionalisms cut across all eras of barn-building traditions, with new regionalisms replacing older ones, but at a different pace in different areas. Barn building traditions were subject to the same basic dynamics as everything else in the world. Nothing stays the same.

Regionalisms might be seen as "a process in the background," to quote Carl Jung. However they performed their magic, they exerted a profound impact on the appearance of barns.

A hundred other regionalisms could be listed here. If, 150 years ago, we could go back and give such a list to farmers and perhaps especially the builders of these barns, we might hear great gasps of amazement. Maybe these farmers and builders influenced other farmers and builders in ways even they were unaware of. They were simply doing their own thing, in their own time.

# Chapter Twelve.
# How to Date a Barn

The Hawbaker stone Sweitzer dated 1813 in Franklin County has several identifiable aspects that reflect a construction date in roughly the second decade of the nineteenth century.

One single aspect of a historic building interests people more than any other characteristic: its construction date. An entire book could be devoted to this subject. So many nuances exist across eras, styles, types, and subtypes that the discussion would require dozens of pages of explanation and illustrations. The following information offers a broad overview.

## Science of Dating Barns

The science of barn dating has advanced considerably since the Dornbusch-Heyl book was published. Establishing a building date involves two essential things. The first is to examine dozens of architectural details in hundreds of barns in many different locations. The second is to systematically study barns that have dates carved into their walls, listing and comparing the design elements that appeared at certain times.

A tracing of a timber with name and date—JACOB STAUFFER - ANNO 1735. This beam as a girt was recycled into a circa 1850 brick Standard barn in Washington Township in Berks County. The timber came from a Ground barn on the Stauffer place.

When I attempt to date a barn that has no carved date, I look at many features to establish its general era of construction. In truth, I often use my intuition. Some barns have lost many of their original features, and the more original features a barn has lost, the more intuition is required. However, no matter what the level of originality, intuition is always a good tool.

When a barn is in close to original condition, it is fairly easy to determine the construction date within ten to fifteen years of its actual date. Such efforts are accurate about ninety percent of the time. For example, I estimated that a Dutch-American three-aisle barn in Somerset County, New Jersey, was built between 1800 and 1820, likely close to 1810. When it was torn down, dendrochronolgy determined that the barn timbers were felled in 1810. Normally this signifies that the barn was built one or two years later, 1811 or 1812. In another case, I visited the Hawbaker stone-to-peak Sweitzer in Franklin County and determined it was built about 1820. Later, I saw a date stone of 1813. However, such estimates are not always so close to the original.

A stone-to-peak Sweitzer dated 1803 in Manheim Township, Lancaster County, occupies a Long family property.

A rare 1795 dated stone-to-peak Sweitzer is located east of Ephrata Borough in Lancaster County. This barn's steep roofline generally reflects this era of construction.

The only eighteenth-century Standard barn ever identified is at the DuPortail homestead in Tredyffrin Township, Chester County. The dated 1792 barn has a steep roof typical of that era.

A stone-to-peak Standard barn in Birdsboro, Berks County, is constructed with brownstone. The roofline is somewhat steep.

A stone-to-peak Standard barn dated 1804 in Bowers, Berks County, destroyed by a tornado in 1998, was immense at 106 feet long. It is the earliest known Standard barn that had a *liegender Dachstuhl*. (Courtesy of Robert Ensminger)

## Dated Barns in Pennsylvania

Five to seven percent of barns in the hearth area have carved or painted dates, usually on the exterior. In their absence, many well-intentioned barn owners believe that a deed date corresponds to the date of both house and barn. Nothing could be further from the truth. Determining a construction date requires the services of a professional barn historian.

By October 2015, I had compiled a list of 195 barns

The Schreiner stone Sweitzer east of Lancaster City was built in 1828. Its proportions are typical of barns built in the first few decades of the nineteenth century. The barn has short vertical slits.

The 1850 Reed stone Standard barn is near the Lebanon–Berks County line near Womelsdorf. Barns in the western half of the hearth often have vertical slits, which are unusual in barns of this era in the eastern half.

The Laubach stone Standard 1834 barn in Lower Saucon, Northampton County, has a rare stone floor at the front of the stable wall. It has a unique mow-stead wall with six doors, all original.

with dates carved or written somewhere on the barns. Most of the barns are in the hearth area. A few of the dates are in question for various reasons. Some of the dates are not actual construction dates, as they were later scribbled on the mow-stead walls and granary room boards. The following decades are represented: 1730s (two), 1740s (two), 1750s (two), 1760s (six), 1770s (two), 1780s (seven), 1790s (twenty-one), 1800s (twenty-nine), 1810s (twenty-five), 1820s (twenty), 1830s (fifteen), 1840s (eleven), 1850s (nineteen), 1860s (eight), 1870s (twelve), 1880s (four), 1890s (two), and 1900s (two).

Although obviously this list is not exhaustive, it offers a general sense of the eras of dated barns. It wasn't until the 1780s that barns began to include, with some regularity, their construction dates. This practice increased in the 1790s and prevailed until the 1830s. Ninety-five barns were seen in the forty-year period 1790 to 1830. This may have been the period when farmers took the greatest pride in their barns and stamped them with dates.

The list of 195 barns includes 130 to 150 barns with authentic dates of construction. From this, a tremendous amount of information may be gleaned of the types of architectural elements and various details that appear in certain barns at certain times. What is needed in the future, however, is an extensive database on all the important diagnostic features of barns across as many eras as possible to help determine dates of construction.

## Standard Barn Style as a Basis for Dating Barns

The Standard forebay barn offers the best example of how a barn may be dated, as it has by far the most date numbers in the hearth area. Only one Standard barn has ever been identified that pre-dates 1800. So the outline to follow only concerns barns in the 1800 to 1920 era. It is assumed that the construction elements in the list are original to the barn being evaluated.

To be certain, other barn types may be dated but they do not offer the clear-cut aspects that exist in Standard barns, and some of these aspects may be applied to other two-level barns. By reviewing this list, barn observers may become increasingly conscious of the characteristics they need to address when dating Standard barns:

- General roof pitches
- End wall construction above the foundation level
- Types of end wall openings
- Roof support construction type
- General use of hewn versus mill-sawn common rafters
- Queen post disposition
- Use of nails in mow-stead walls and granary walls
- General use of wood species in mow-stead walls
- Boards with parallel saw marks versus circular saw marks

## Barns with Stone-to-Peak (SP)

The earliest Standard barns almost always had stone-to-peak on their end walls. They also have vertical loopholes or slits on their end walls and rear walls. The slits are positioned in various arrangements and normally appear in three or four rows. A few rare barns have five rows of slits, and one exceptional barn just northwest of Womelsdorf in Berks County has six rows of slits. In general, slits vary from about 18 to 36 inches high. Barns with slits were quite commonly built until circa 1830 and occasionally beyond. In the western counties in the hearth, slits appear in barns as late as the 1850s.

Barns with slits have attendant early features in their interiors. Especially in the eastern counties, they often have Principal Rafter Systems. It may be said that vertical slits and Principal Rafters frequently go hand-in-hand. When early Standard barns have Common Rafter Systems, they usually have *stehender Dachstuhl* roof support. These are basically vertical queen posts that support the purlin plates. Most common rafters in stone-to-peak Standard barns are hewn.

Many of these barns have mow-stead walls with early style nails. The nails may be of the wrought variety (not common in early Standard barns) with shafts that have square cross sections and pointy ends. Such nails are seen up until about 1810 and occasionally beyond. Transitional nails that often have handmade or wrought heads have machine-made stamped shafts. These nails are seen in the 1800 to 1820 era and beyond. Cut nails or fully machine-made ones are considered late vintage. They have square heads and rectangular cross sections. These nails generally first appear about 1820 and were in use many decades after this. Earlier versions of cut nails did appear prior to 1800.

Another dating feature is mow-stead walls with horizontal oak boards. Oak is seen up to about 1820. Pine is considerably more prevalent in barns built after 1820. Tulipwood or poplar was also occasionally used.

Most stone-to-peak Standard barns and many Sweitzers built in the 1780 to 1830 or 1840 era have vertical slits. Many barns with slits in the west half of the hearth continued to be built ten to twenty years beyond that.

Many Sweitzers in the 1770 to 1810 era, and some Standard barns in the 1805 to 1840 or even 1850 era were built with *liegender Dachstuhl* roof structures. Some Ground barns in the late 1700s have this roof type. (Courtesy of Robert Ensminger)

Occasionally some barns built after 1820 or 1830 have mow-stead walls with poplar boards. These boards are wide, up to 18 inches.

The great majority of barns built after 1810 or 1820 have pine mow-stead wallboards. Boards in pre-1800 or 1810 barns are often secured with wrought nails. In the 1800 to 1830 era barns have boards secured with transitional (partially wrought) style nails. In post-1820 or 1830 barns, boards are secured with square-headed cut nails.

The earliest Standard barns have the finest workmanship, including finely executed timber joinery. As the years went by in the nineteenth century, craftsmanship declined in quality. Many barns have to be seen to obtain a sense of the various levels of excellence.

## Barns with Louvered Windows

Stone-to-peak Standard barns built in the next "dating era" include louvered windows (LW) on their end walls and rear walls. These barns were built to some degree in the 1810s but were more prevalent in the 1820s and 1830s. Not many of these barns were built after 1840 or so. Standard barns with louvered windows came onto landscapes on average ten to fifteen years after barns with vertical slits.

These barns may be a little bigger than barns with vertical slits. As the decades progressed in the nineteenth century, there was a need for bigger barns. Quite often, but to a lesser extent than barns with slits, stone-to-eaves-level structures with LW also have Principal

Some barns built as early as 1800 have louvered windows. Such windows are commonly seen in post-1820 or 1830 barns until 1840 or perhaps 1850.

Rafter Systems. But the later built ones start to employ Common Rafter Systems. In this case, most of these barns have either the earlier style vertical queen posts or, less often, the later style canted queen posts.

## Barns with Stone-to-Eaves-Level

As the nineteenth century progressed, Standard barns started to have less stone in their end walls. Perhaps beginning about 1820 and certainly by 1830, some Standard barns were being built with stone only to the eaves level (SEL) on their end walls. It was likely not until about 1840 that these barns started to proliferate. They were built into the 1860s and sometimes into the 1870s. Both stone to peak barns and SEL barns were being built at the same time, in the 1830 to 1850 era.

Relatively few SEL barns with vertical slits were built. Barns with louvered windows were somewhat more prevalent.

It seems that not many SEL barns were built with Principal Rafter Systems. When they were, Principals were included at their end walls. Common Rafter Systems were more common in SEL barns than in stone-to-peak barns. The earlier 1830 to 1850 SEL barns tend to have more vertical queen posts. Later ones have canted queen posts.

A few of the earliest SEL barns have transitional-era nails in their mow-stead walls and granary boards. But most of these barns have square headed cut nails in their wood boards. Most of the mow-stead walls have pine boards and occasionally tulipwood (yellow poplar) boards.

Stone-to-eaves-level Standard barns such as this South Whitehall Township barn were built in the 1840 to 1870 era. These barns quite infrequently have vertical slits or louvered windows.

## Other Stone-to-Peak Barns with Wall Openings

Two other end-wall-opening types are prevalent, both of them in the greater Lehigh and Saucon Valleys. One type has circular loopholes on the end walls. They are arranged in four or five rows on the end walls and two rows on the rear walls. The holes, about 10 inches in diameter, are outlined with ten or eleven radiating bricks. These barns are strictly limited to the Saucon Valley in the hearth area. They rarely, if ever, occur west of Bethlehem, and they do extend out to the Delaware River. The earliest ones occured about 1810 to 1820 and the latest ones in the 1840s. Some of the barns have Principal Rafter Systems and later ones have Common Rafter Systems. Some of these earlier barns have transitional-style nails and later ones have square-headed cut nails. In addition, when Principal Rafter Systems are lacking, the earlier barns have vertical queen posts while the later ones have canted queen posts.

A few stone Sweitzers and some Standard barns within eight to ten miles of the Delaware River in the Saucon Valley have circular loopholes; they also appear in northern Bucks County. They were built 1810 to 1850.

The other barn type has brick inset pattern openings on the end walls and rear walls. They are either diamond shaped or abbreviated "sheaves of wheat." They are essentially restricted to the greater Lehigh River Valley. Basically, statements about dating parameters seen in the barns with circular brick lined loopholes apply to the barns with inset brick patterns. The only exception concerns roof structures where Principal Rafter Systems are not common.

## Stone-to-Peak Barns without Wall Openings

Many SP Standard barns have no wall openings above the basement. The only exception are barns with house-size windows just below the peak on their end walls; only a few of the windows are likely original.

These barns were likely built in the 1820s and became more common after the 1840s. Beyond about 1860, the barns were not at all commonplace. I have not seen any SP barns without wall openings dating from the 1870s, and just two dated SP barns built in the 1880s.

Many Standard barns in the greater Lehigh River Valley and somewhat beyond have diamond-shaped brick pattern insets. They were built from 1820 or 1830 to about 1860.

Numerous stone-to-peak Standard barns were built in the 1830 or 1840 to 1860 era with no distinctive wall openings. A few such barns were built into the 1870s and even the 1880s.

A fair number of barns in the hearth were built with varying degrees of brick; most were built 1830 to 1870.

## Brick Standard Barns

Brick Standard barns were infrequently built before about 1830 and not often in the 1830 to 1840 era; they appear sporadically in the hearth area. They were predominantly built between the 1840s and 1870s. Many of the brick barns have date-stones, so there is no mystery surrounding their date of origin.

Two dozen or so brick barns are known in northern Montgomery County, and a few concentrations are seen in Lancaster County. They are more common west of the Susquehanna River, especially in Franklin County.

Most of these brick barns have Common Rafter Systems. The earlier ones have vertical queen posts, and after 1840 to 1850, canted queen posts are more prevalent. Likely, a majority of these barns had milled rafters while earlier ones had hewn rafters.

An exception is a barn built circa 1850 or 1860 south of Bally in Berks County, which has a Principal Rafter System. Mow-stead wall boards are usually secured with square headed cut nails.

## Frame Standard Barns

Frame barns were built continuously from 1840 to 1920 and were less labor-intensive than stone barns.

Some of those built in the 1840s or 1850s had German roof structures, and an equal number of barns had Principal Rafter Systems. A number of the frame barns have Common Rafter Systems with hewn rafters. One dated in the early 1850s was in the Shartlesville area of northwestern Berks County. Likely a great majority of these barns had canted queen posts. Most of the 1840 to 1890 barns have mow-stead walls and granary walls with boards secured with square headed cut nails. After 1890 or 1900 the barns have round headed wire nails. In many post-1870 barns, boards began to have circular saw marks rather than the earlier parallel saw marks.

Frame Standard barns are quite common in many areas of the hearth. Few were built prior to about 1830 but they proliferated in the 1850 to 1910 time frame.

## Dendrochronology

Dendrochronology is the science of tree-ring dating. It is often a last resort after all other dating efforts have failed. Either transverse slices or cores of wood are obtained from timbers of buildings. Usually the wood samples are oak, which was the species of choice for so many builders in the eighteenth and nineteenth centuries in southeast Pennsylvania.

The slices or cores are prepared (sanded) and then sent to a certified laboratory.

The number of annual growth rings is counted and certain growth patterns are detected and compared with so-called master lists. If the testing is successful, felling dates of trees that the timbers came from are established. The dates of the various samples ideally should coincide. It is then assumed that the timbers were used to make the building the following year or possibly the second year thereafter. Perhaps eighty percent of the time, dendro-dating timbers accurately establishes a construction date.

One example of successful dendro-dating involved a frame Standard barn in Upper Frederick Township, Montgomery County that I had dendro-dated. It checked out that timbers were felled in 1861. Thus the barn was likely erected in 1862 or 1863. This coincides with some of the barn's construction features. The science of dendro-dating works.

Many barns have been dendro-dated in the eastern United States in the past quarter of a century. The only substantial drawback to this procedure is the cost, which is often prohibitive. Dendro-dating is the ultimate means in knowing almost the exact dates of construction when a building lacks a date-stone.

In the future, perhaps a wide database will be established for barns across their range of building eras. This will greatly aid us in knowing dates of barns. The full gamut of dendro-dating should include remnant timbers thought to date from quite early barns, pre-dating 1780 or so.

Dendro-dating of timber frame barns can indicate the age of construction. Oak was a common tree species used. This oak slice (above right and lower left), from a large oak in the Oley Valley in Berks County, shows annual growth rings.

## Summary

The determination of dates of construction can be both an exciting and challenging pursuit. As Eric Sloane said in his 1966 book *An Age of Barns,* "Dating an old, unrecorded barn is a job of detection. Even then, the answer is usually nothing more than an approximate guess and must contain the word circa." His assertion is still largely correct, though it is now a bit more refined than that, but the inclusion of the qualifier "circa" has merit.

Sorting out all the elements of design, fabric, and construction is an arduous task indeed. Sometimes conflicting aspects of barns present themselves. Dating is essentially a balancing act. Sometimes dating a barn almost seems like a compromise. It is really a matter of observing many barns until the wide array of features that individual barns possess are sorted out and things become intelligble.

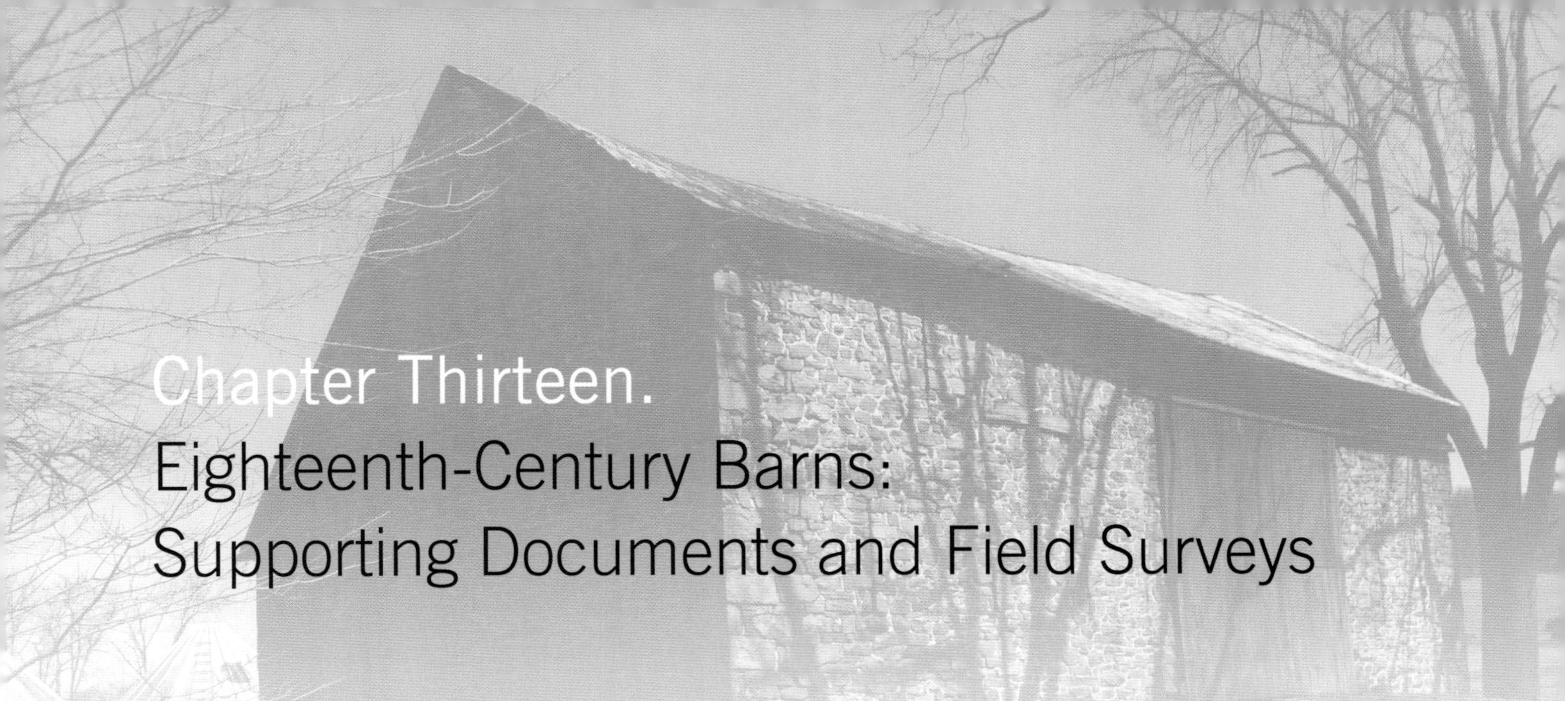

# Chapter Thirteen. Eighteenth-Century Barns: Supporting Documents and Field Surveys

Barns were built in great numbers from the early 1700s to circa 1900, almost all of them rural vernacular structures. A remarkably small number have survived from the earliest years of this nearly 200-year-long period. Many of these early barns were supplanted by more efficient structures in response to new economic and agricultural conditions. This led to tremendous attrition of earlier barns. Only about one dozen barns or remnant barns pre-date 1775. Prior to about 1975, little documentation was done on barns in the hearth.

With this in mind, how does one obtain even a fair idea of what barns looked like in the hearth area in the eighteenth century? The answer may be found in principally two areas of inquiry. Documents of whatever kind they may be, including oral accounts and photographs, serve as the only source of information on these barns. The second involves surveying surviving barns.

## Federal Direct Tax of 1798

Prior to about 1850, documented barns include only written accounts, sketches, and paintings. After 1850, photographs were taken, though few interior views are available. One of the best documents of the eighteenth century is the Federal Direct Tax of 1798. The federal government conducted an extensive survey of buildings of all types in twelve states. These structures were measured and categorized for building type, materials usage, and dimensions. Of the three categories, dimensions are only represented some of the time. Land owners were taxed using criteria based on these aspects. This was to pay for an impending war with France. By this means, tens of thousands of buildings were recorded to varying degrees.

Hearth-area counties listed in the 1798 tax include Philadelphia, Bucks, Montgomery, Northampton, Delaware, Chester, Lancaster, and Dauphin. Gaps exist in some areas, including nearly all of Berks County. The eight counties represent nearly 16,000 barns. One can only imagine the time and energy expended to survey all the homesteads and other properties that yielded information on such a number of buildings. And recall that this was at a time when transportation was slow.

Barns listed in the several counties where they occurred include citations to a few materials and often combinations of materials that were used. The materials were log, stone, and frame, but only rarely brick. In certain cases, the number of floor levels is also recorded. However, this data has to be interpreted carefully, because the words "first" and "second" in the barn descriptions do not necessarily indicate the number of stories as we would define them today. Certain Ground barns had stone for the first several feet in their end bays and the walls above were either log or frame. Today, some people may interpret this as two stories when in fact the barn may have had just one story. So the actual meaning of first and second requires clarification, but no one is alive from the eighteenth century who might help us in this regard. It seems that the building assessors did not record the precise style or type of construction.

Barns in the Direct Tax list represented a broad range of ages. Some of the barns are referred to as "old," and less frequently "very old." These are, of course, relative terms. What one assessor regarded as "old" another might call "very old." From this information then, we can know the age of the barns only in a general way.

The stone and log Ground barn in the Argus area of northern Bucks County is of late-eighteenth-century vintage. Some Sweitzers of this era in the hearth are stone and log.

By 1798, many first-generation barns had been replaced with second-generation structures. It is possible that certain second-generation barns were actually older than some first-generation barns. So we can't tell the exact nature of the first- or second-generation barns recorded.

Some township surveys listed the dimensions of every barn, while other townships included the dimensions of only a few. This may be a result of different surveyors, the need to complete the work before the onset of winter, or inaccessibility to the barn perimeters.

## Benefit of 1798 Federal Direct Tax

Buried deep within the Federal Direct Tax data, certain arcane truths about eighteenth-century barns exist. To ferret out these things, various pieces of information must be looked at critically. For example, farms with certain acreages may include barns built with particular construction materials. A farm with many acres, perhaps 150 or more, may have had stone barns, while a farm with considerably less acreage may have had the considerably cheaper log barns, or more rarely, barns built of frame. Farms with extensive acreage, often have considerably more tillable acres and therefore the successful farmer could likely afford to build substantially more expensive stone barns. A comparison of dozens or even hundreds of homesteads may yield statistically important facts about particular dynamics involving barn-building traditions. Another correlation may emerge, such as the relationship of tillable acres to the size of barns. This is an idealized manner of looking at things, as it does not take into account other potential factors such as availability of stone in particular areas. Suffice to say, the Federal Tax list may well be a heretofore unrecognized treasure trove of information.

## Other Eighteenth-Century Documents

The 1798 Federal Direct Tax will remain the single greatest eighteenth-century documentary source available to barn researchers. But other documents exist, however rare they may be. They may surface at any turn, in any source, at any time. One example, cited in chapter nine, concerned a barn that burned in September 1760 on a Roberts family property in the Pencoyd area of Lower Merion Township, Montgomery County. Algeron Roberts in 1791 erected another barn made of stone, and the Religious Society of Friends censured him for such extravagance. These sorts of documents lend extraordinary insight into the dynamics, in this example, of local Quaker barn-building traditions. Whatever those customs exactly entailed is left for future historians.

Such information sources are hit or miss, as they are not indexed in any archival resource. However, one potentially rich and easily accessible source is inventories of estates of the deceased. Often appraisers referred to the number of bushels of farm crops "in the barn," with the corresponding values. We never know what insight these arcane citations may offer.

## County-Wide Barn Surveys

Another resource that could be developed is a road-by-road survey of extant barns in the hearth. Not all barns would need to be entered, especially Standard barns, as all of them, except the Duportail barn, are from the nineteenth century and the first two decades or so of the twentieth century. However, any barn suspected of pre-dating 1800, such as many Sweitzers and certain stone Ground barns, should be examined for an approximate construction date. The area of highest concentration of Sweitzer barns is the northern two-thirds of Lancaster County, the southern half of Lebanon County, and the western half of Berks County. Such an undertaking covers approximately 1,500 square miles and would yield perhaps a few dozen pre-1800 barns.

In the hearth area I have located thirty eighteenth-century barns or remnants of barns, out of the approximately 5,000 to 6,000 barns examined since 1975. These early barns all have or had dates applied during construction. Perhaps another fifteen to twenty barns of this era, but without dates, have also been recorded. Not all of the dated barns have survived.

The dated 1793 stone Grosser Sweitzer in Montgomery County had recessed forebay end walls. The interior had a variant *liegender Dachstuhl*. The barn was removed around 2000. (Courtesy of Robert Ensminger)

Six dated barns are ground structures—four fully intact ones and two remnants—fourteen are Sweitzers, one is a Standard barn, two are English Lake District barns, one is an ELD barn that was later converted to a Double Decker, and six barns are of a nondescript nature.

## Lehigh County Barn Survey

I also did a survey of all the barns in Lehigh County, where I live. I performed this survey over a forty-six-day period on selected days in February 2012, from January to April 2013, and in June 2013. This was primarily a windshield survey of fifteen townships, five boroughs, one village, and many hamlets. The survey yielded 1,670 barns. A few dozen were entered to record certain details. Any barn suspected of dating around the year 1800 or possibly before was particularly scrutinized to obtain at least a general date of construction.

The results of the county survey produced just one definite eighteenth-century barn, a circa 1780 stone and frame Sweitzer. Three other barns seemed to be circa 1800 barns—two Sweitzers and an English Lake District barn. In addition, a dated 1800 stone and frame Sweitzer was discovered.

## Forebay Barns in the Lehigh County Survey

Several drives through a number of rural areas in Lehigh County will quickly reveal that forebay barns are by far the most prevalent barn type. Indeed, the county-wide survey revealed that sixty-eight percent were forebay barns, totaling 1,140 barns.

The land areas in Lehigh County's fourteen townships vary greatly and therefore the number of barns in each township varies considerably. Lynn Township, the largest, harbors the highest number of forebay barns—138. Five other townships each possess more than one hundred barns. Four townships have between fifty and one hundred barns, and five townships each have forty-five or fewer barns.

In the nineteenth century, the one-level Ground barn was one of the most significant barn categories, but its importance declined as the century advanced. It trailed only the forebay barn in agricultural importance. By comparison, the Lehigh County survey yielded fifty-two Ground barns, or three percent of the total barn population. This signifies that forebay barns outnumbered Ground barns by a factor of twenty-two to one. The predominance of forebay barns likely commenced in the second quarter of the nineteenth century.

Two bar graphs (opposite) show the distribution of barns in various categories in two townships—Upper Macungie (eighty-eight barns) and Lower Macungie (117 barns). The widely divergent number of barns in each category largely reflects the transitional zone between limestone and slate. The southern limestone belt provided excellent stone for construction. The county survey indicated that many more stone barns were built in the southern half than in the northern half.

Other bar graphs (following) show the distribution of materials used in barns documented in the 1798 Direct Tax. The types and forms in each township vary to a great degree. For example, stone barns were more prevalent in the southern townships in 1798.

Barn size and materials varied as builders and farmers alike responded to personal and local circumstances.

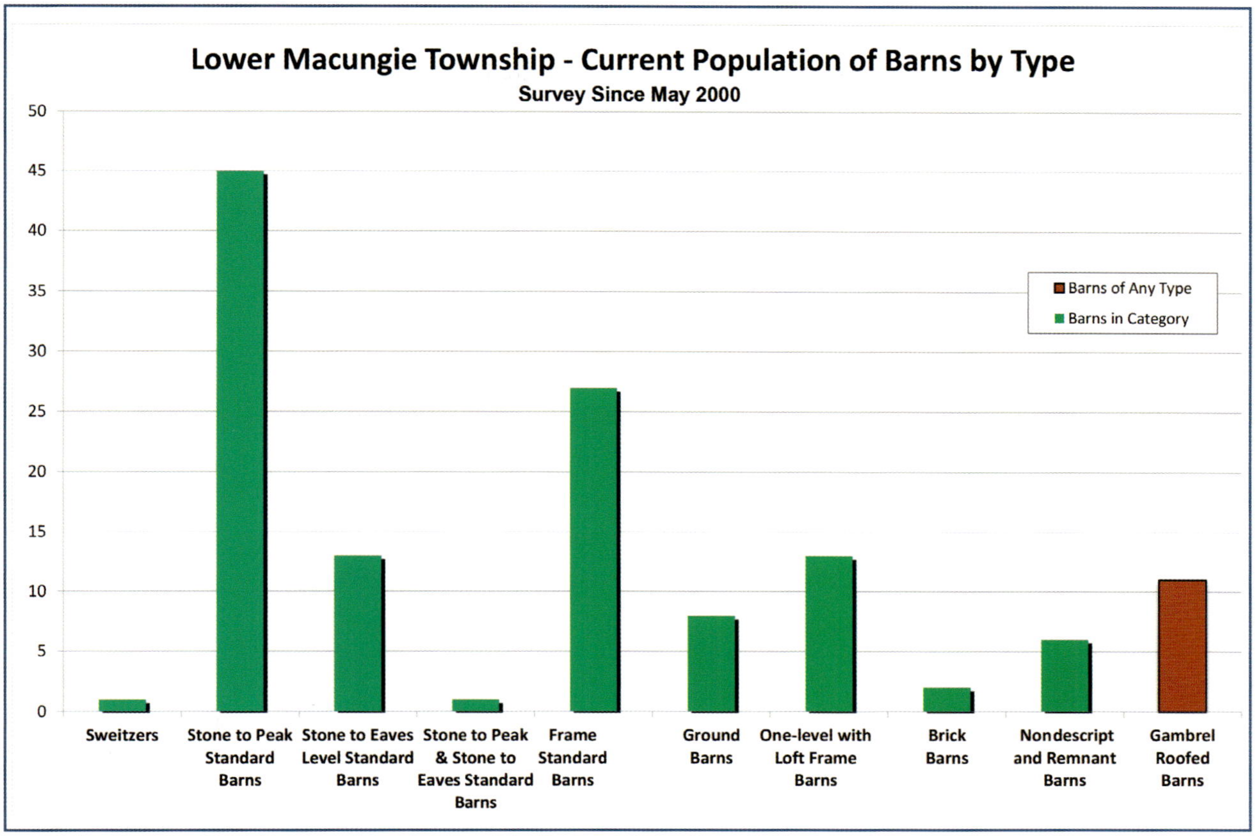
Lower Macungie Township - Current Population of Barns by Type
Survey Since May 2000
50
45
40
35
30
25
20
15
10
5
0
Barns of Any Type
Barns in Category
Sweitzers
Stone to Peak Standard Barns
Stone to Eaves Level Standard Barns
Stone to Peak & Stone to Eaves Standard Barns
Frame Standard Barns
Ground Barns
One-level with Loft Frame Barns
Brick Barns
Nondescript and Remnant Barns
Gambrel Roofed Barns

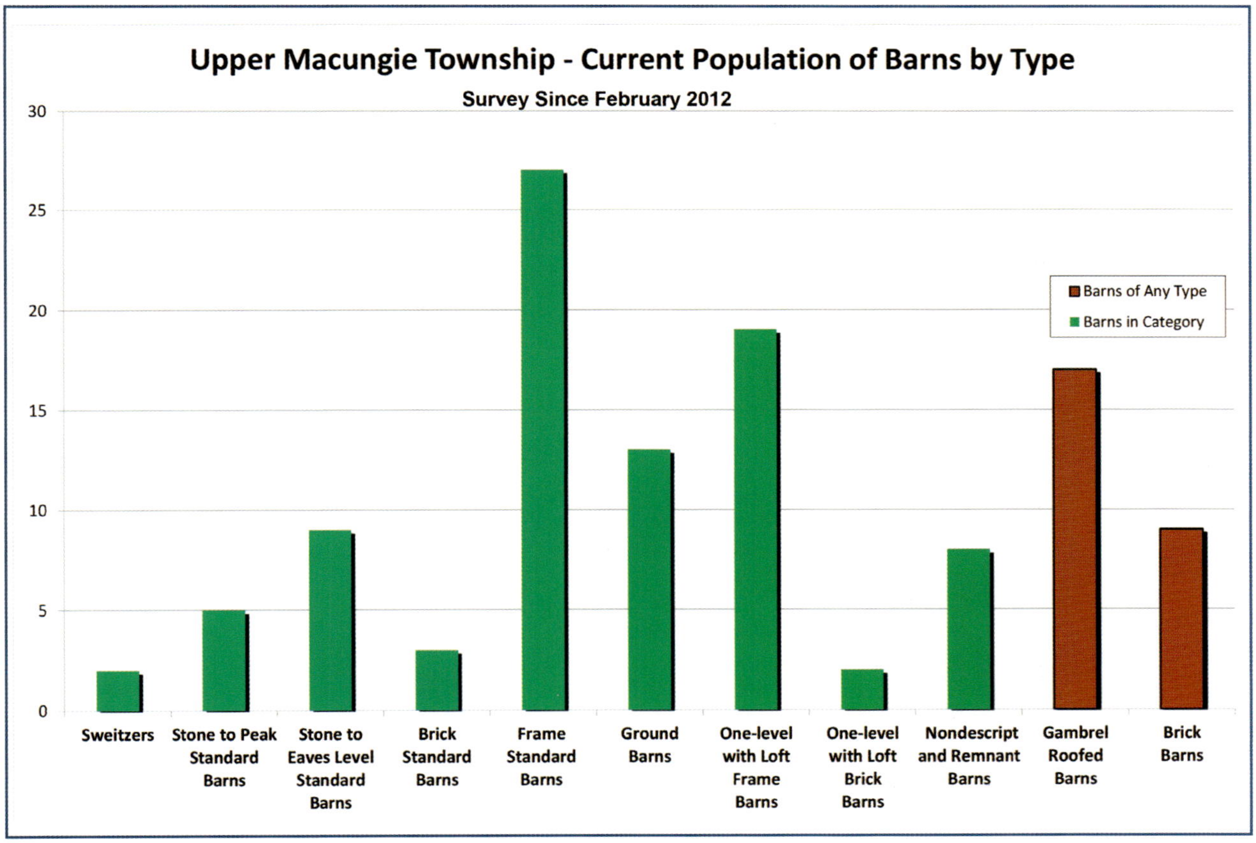
Upper Macungie Township - Current Population of Barns by Type
Survey Since February 2012
30
25
20
15
10
5
0
Barns of Any Type
Barns in Category
Sweitzers
Stone to Peak Standard Barns
Stone to Eaves Level Standard Barns
Brick Standard Barns
Frame Standard Barns
Ground Barns
One-level with Loft Frame Barns
One-level with Loft Brick Barns
Nondescript and Remnant Barns
Gambrel Roofed Barns
Brick Barns

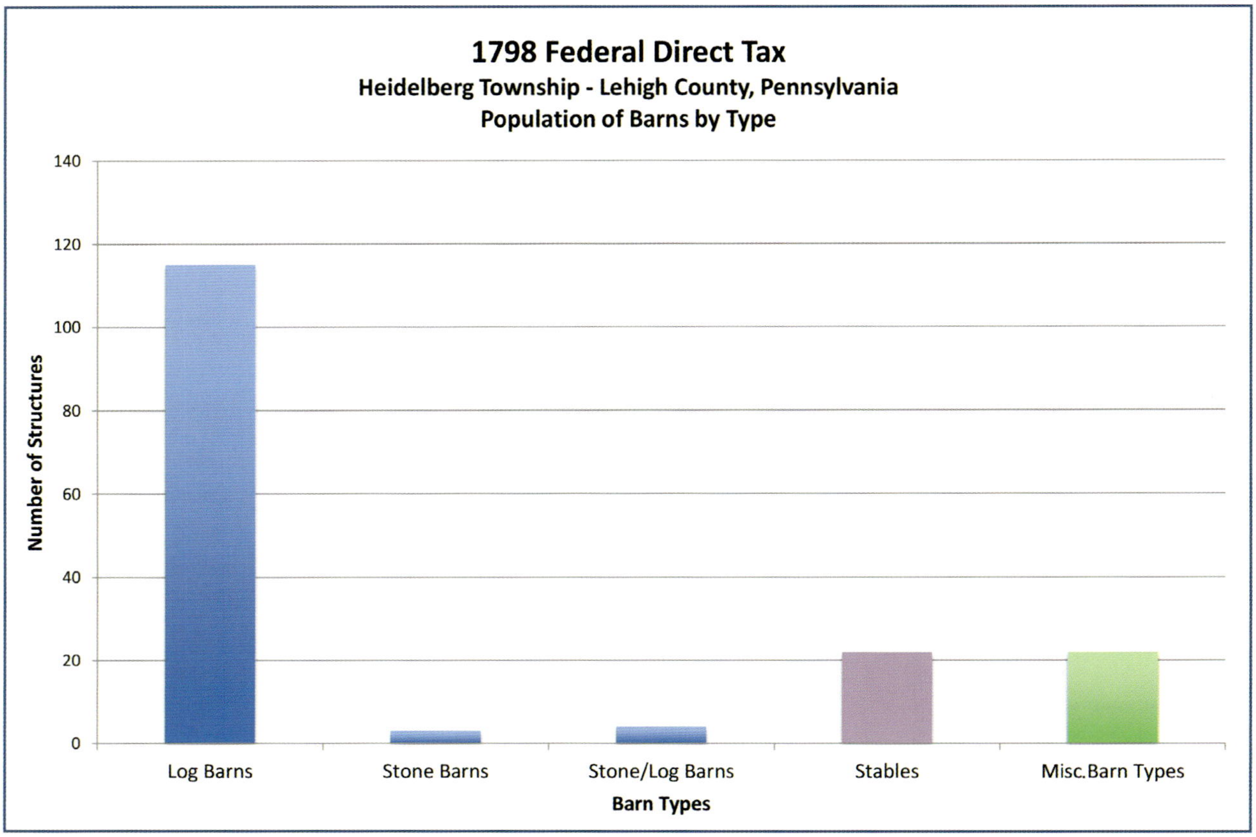
1798 Federal Direct Tax
Heidelberg Township - Lehigh County, Pennsylvania
Population of Barns by Type
Number of Structures
140
120
100
80
60
40
20
0
Log Barns
Stone Barns
Stone/Log Barns
Stables
Misc.Barn Types
Barn Types

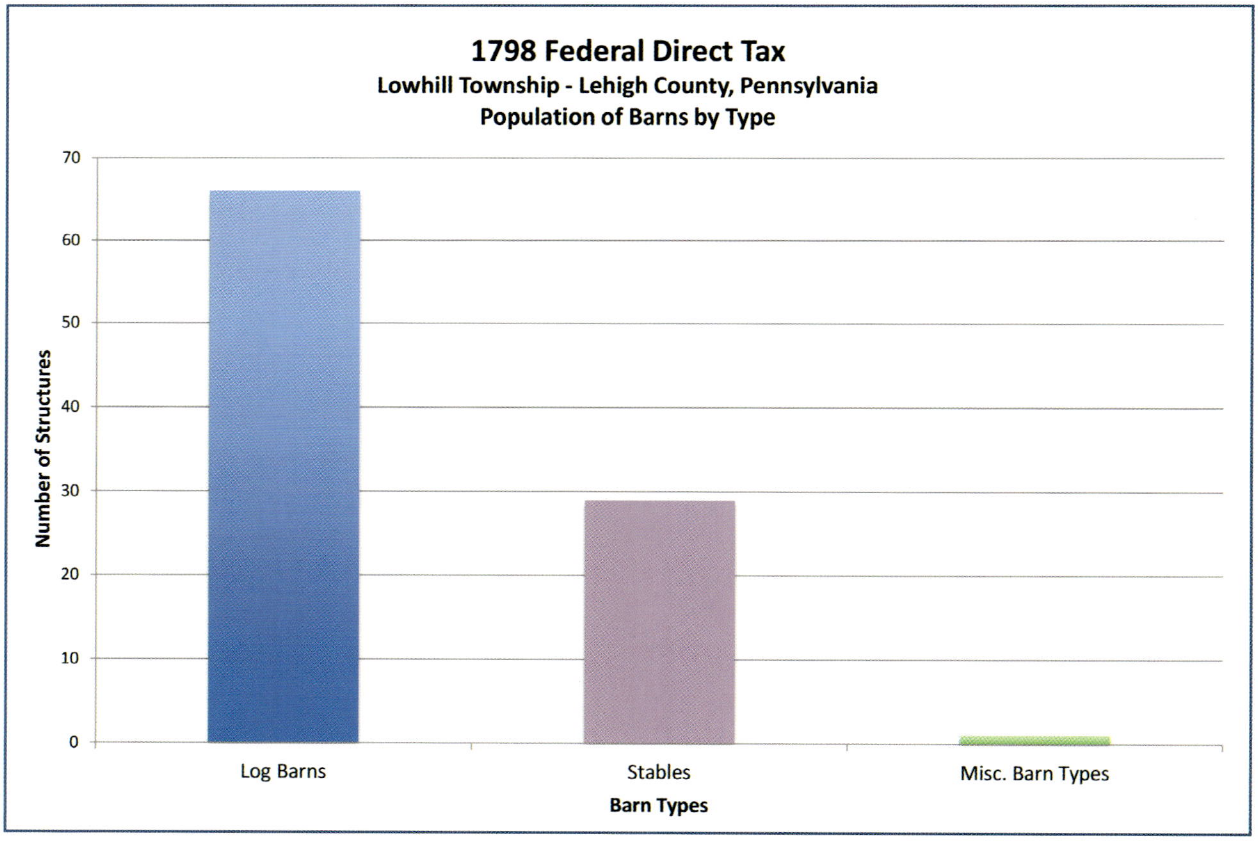
1798 Federal Direct Tax
Lowhill Township - Lehigh County, Pennsylvania
Population of Barns by Type
Number of Structures
70
60
50
40
30
20
10
0
Log Barns
Stables
Misc. Barn Types
Barn Types

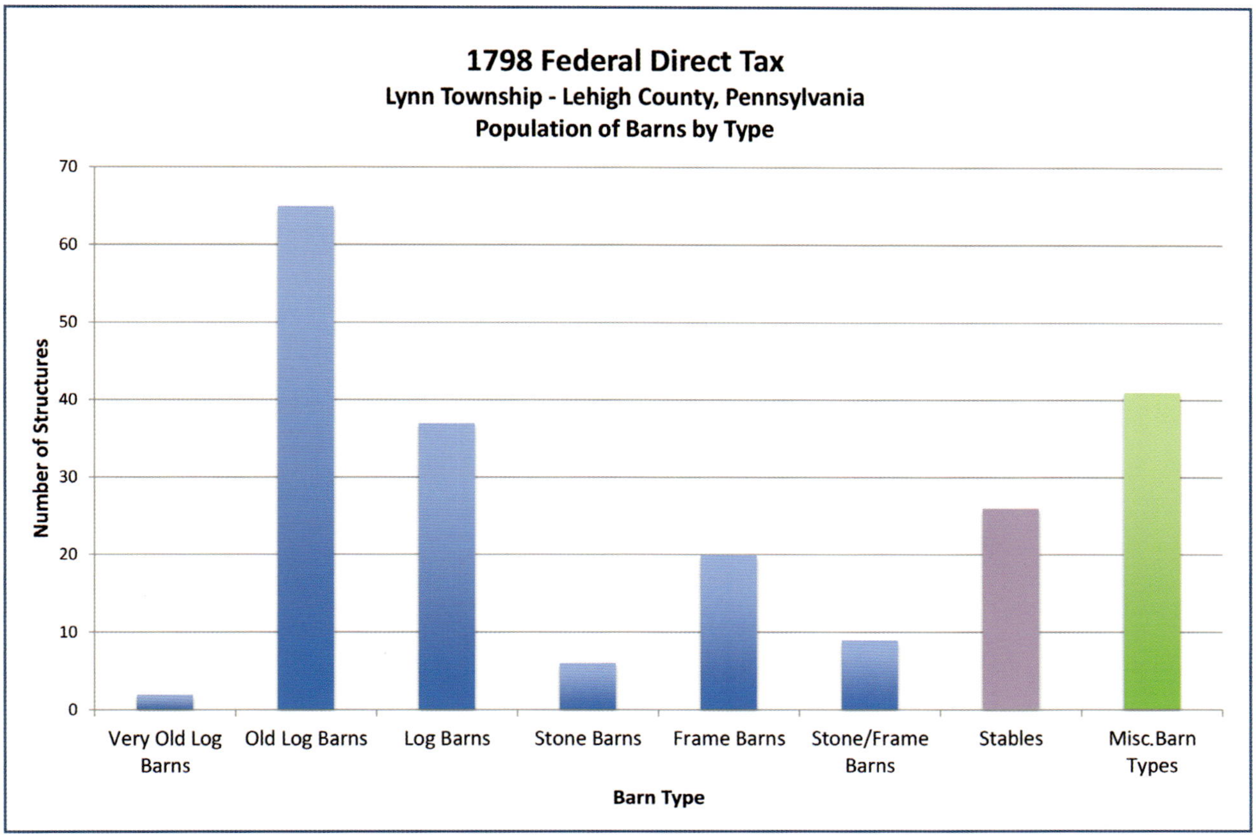
1798 Federal Direct Tax
Lynn Township - Lehigh County, Pennsylvania
Population of Barns by Type
Number of Structures
0
10
20
30
40
50
60
70
Very Old Log Barns
Old Log Barns
Log Barns
Stone Barns
Frame Barns
Stone/Frame Barns
Stables
Misc.Barn Types
Barn Type

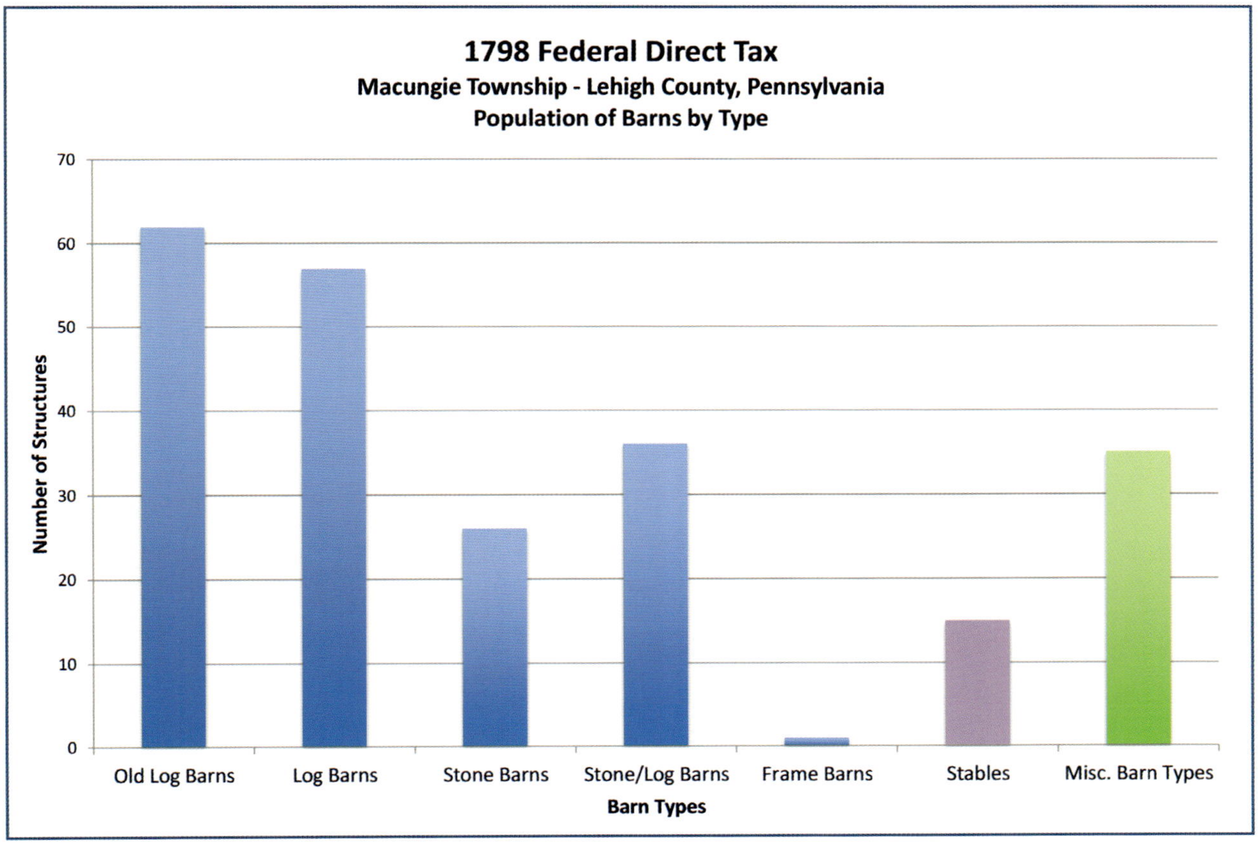
1798 Federal Direct Tax
Macungie Township - Lehigh County, Pennsylvania
Population of Barns by Type
Number of Structures
0
10
20
30
40
50
60
70
Old Log Barns
Log Barns
Stone Barns
Stone/Log Barns
Frame Barns
Stables
Misc. Barn Types
Barn Types

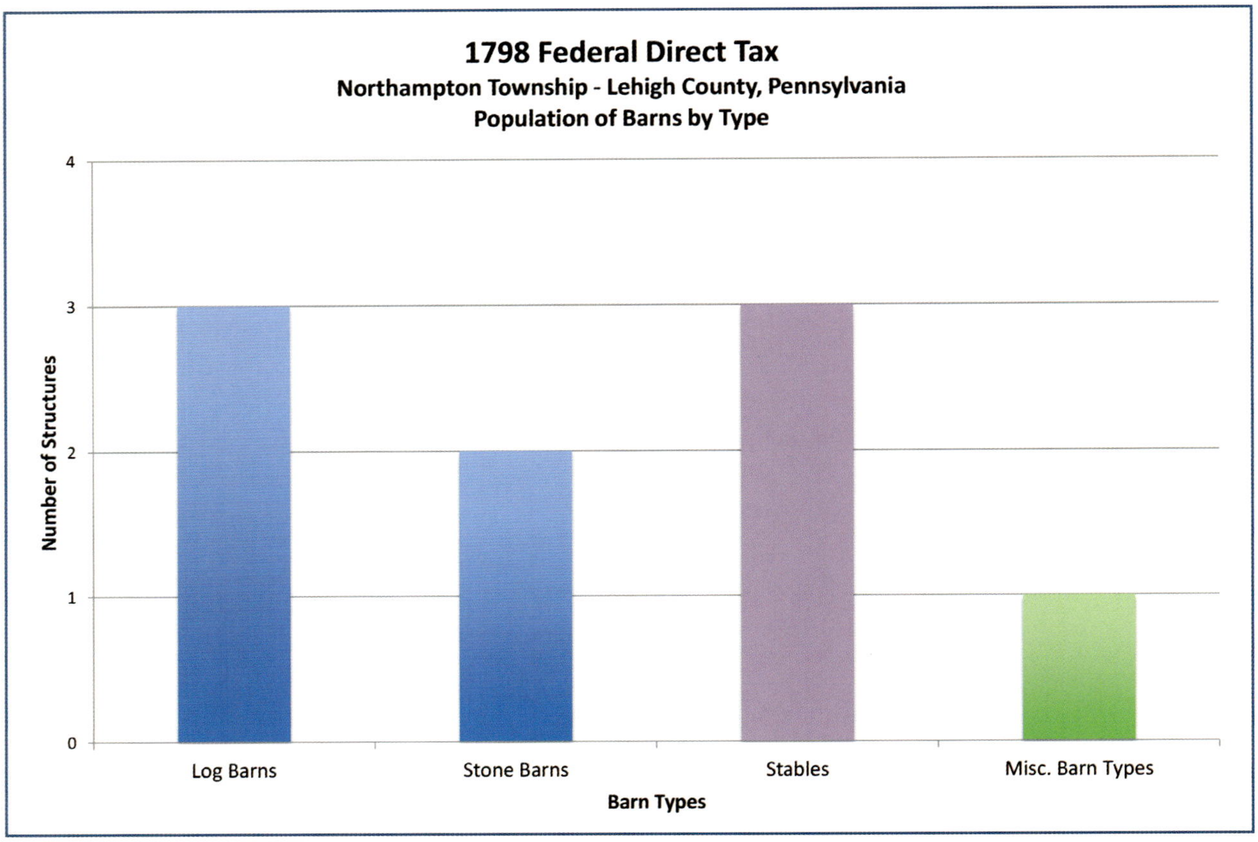
1798 Federal Direct Tax
Northampton Township - Lehigh County, Pennsylvania
Population of Barns by Type
Number of Structures
0
1
2
3
4
Log Barns
Stone Barns
Stables
Misc. Barn Types
Barn Types

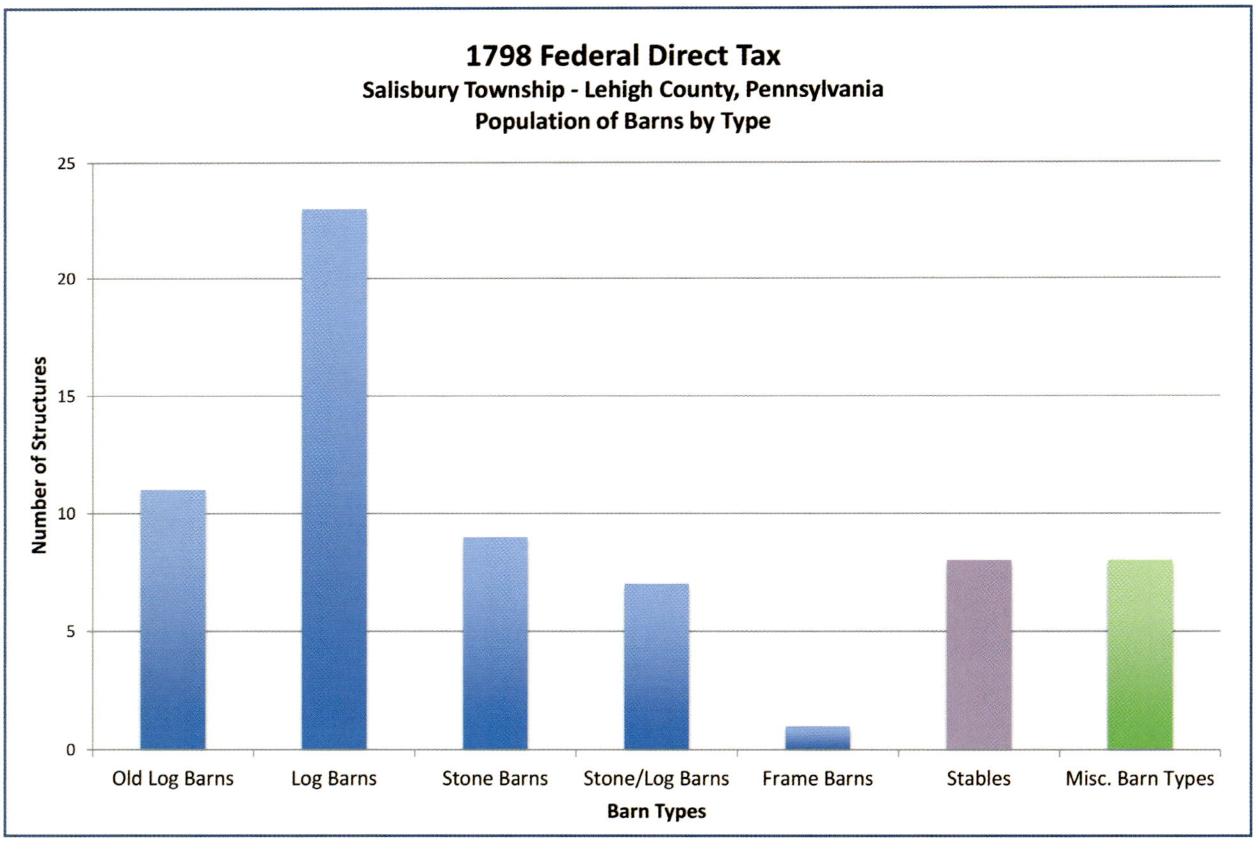
1798 Federal Direct Tax
Salisbury Township - Lehigh County, Pennsylvania
Population of Barns by Type
Number of Structures
0
5
10
15
20
25
Old Log Barns
Log Barns
Stone Barns
Stone/Log Barns
Frame Barns
Stables
Misc. Barn Types
Barn Types

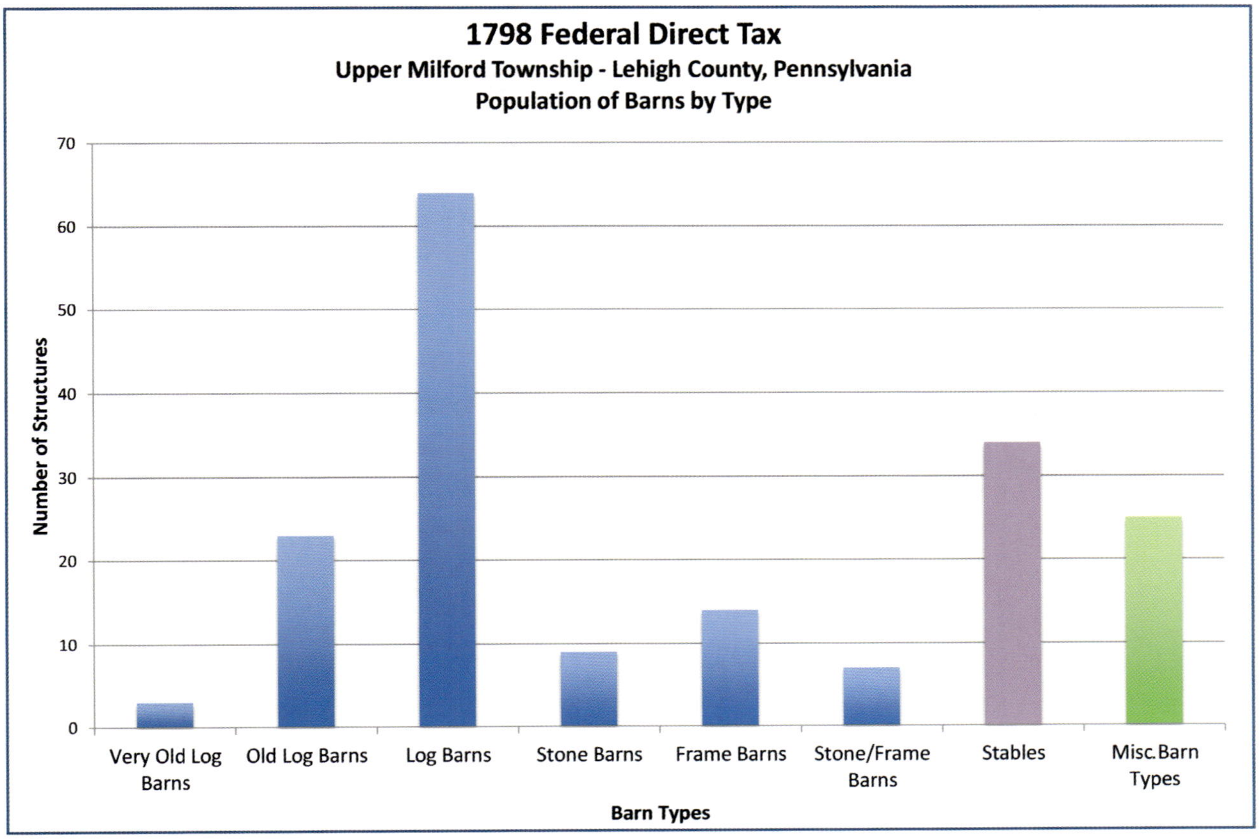
1798 Federal Direct Tax
Upper Milford Township - Lehigh County, Pennsylvania
Population of Barns by Type
Number of Structures
0
10
20
30
40
50
60
70
Very Old Log Barns
Old Log Barns
Log Barns
Stone Barns
Frame Barns
Stone/Frame Barns
Stables
Misc. Barn Types
Barn Types

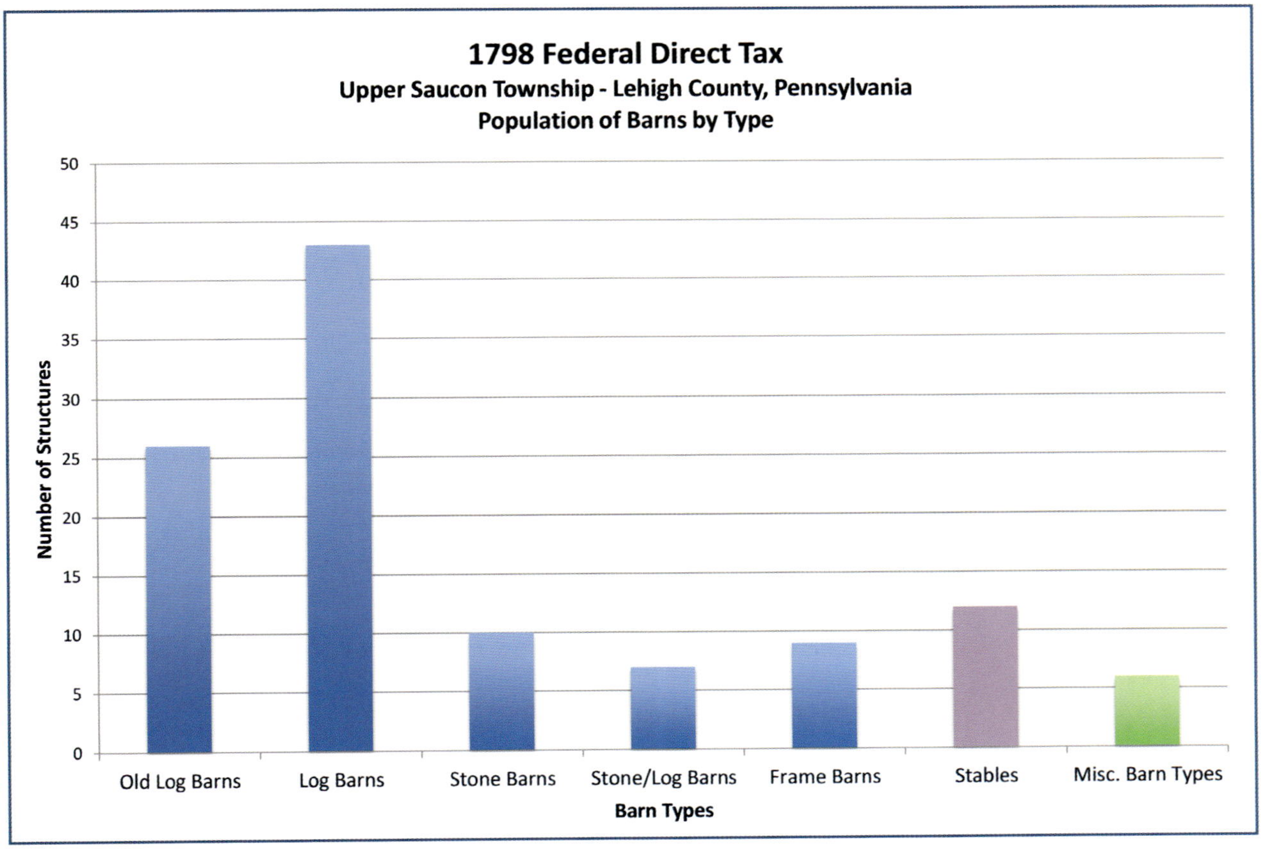
1798 Federal Direct Tax
Upper Saucon Township - Lehigh County, Pennsylvania
Population of Barns by Type
Number of Structures
0
5
10
15
20
25
30
35
40
45
50
Old Log Barns
Log Barns
Stone Barns
Stone/Log Barns
Frame Barns
Stables
Misc. Barn Types
Barn Types

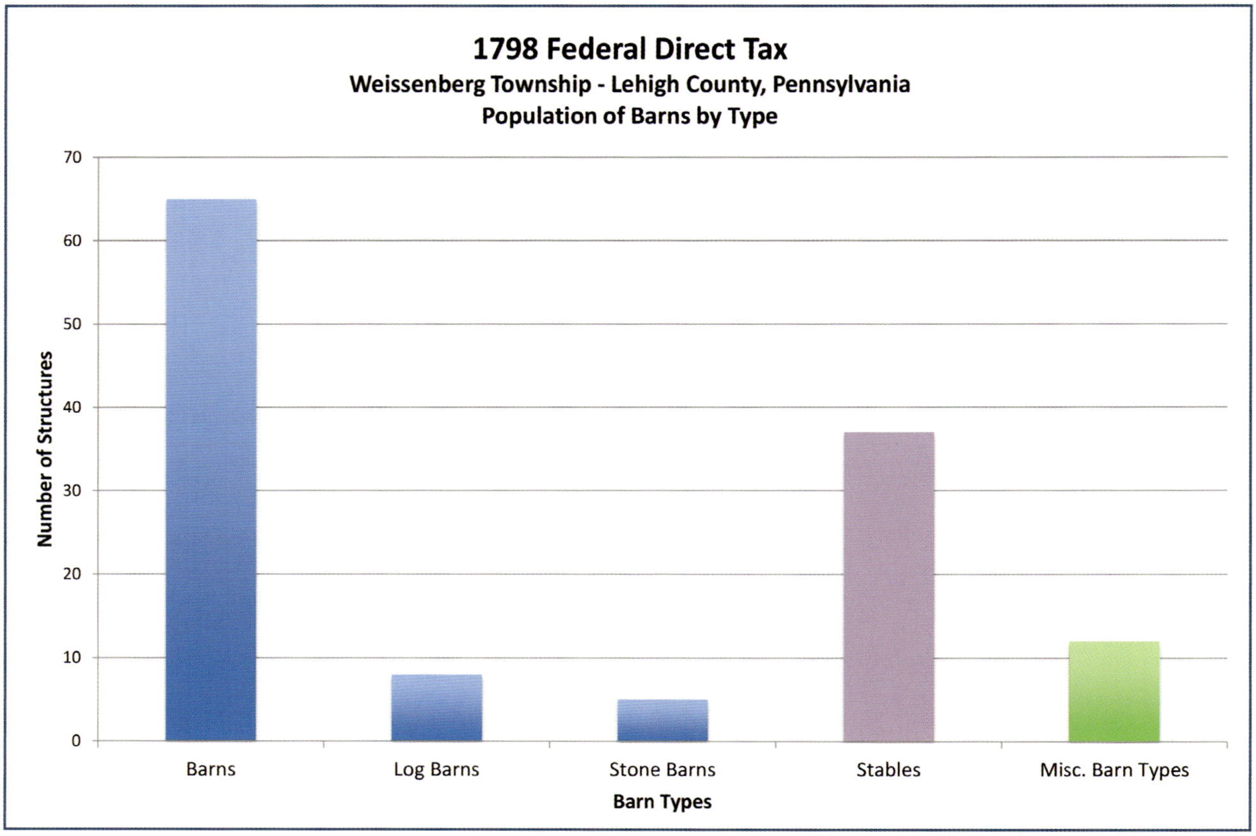
1798 Federal Direct Tax
Weissenberg Township - Lehigh County, Pennsylvania
Population of Barns by Type
Number of Structures
0
10
20
30
40
50
60
70
Barns
Log Barns
Stone Barns
Stables
Misc. Barn Types
Barn Types

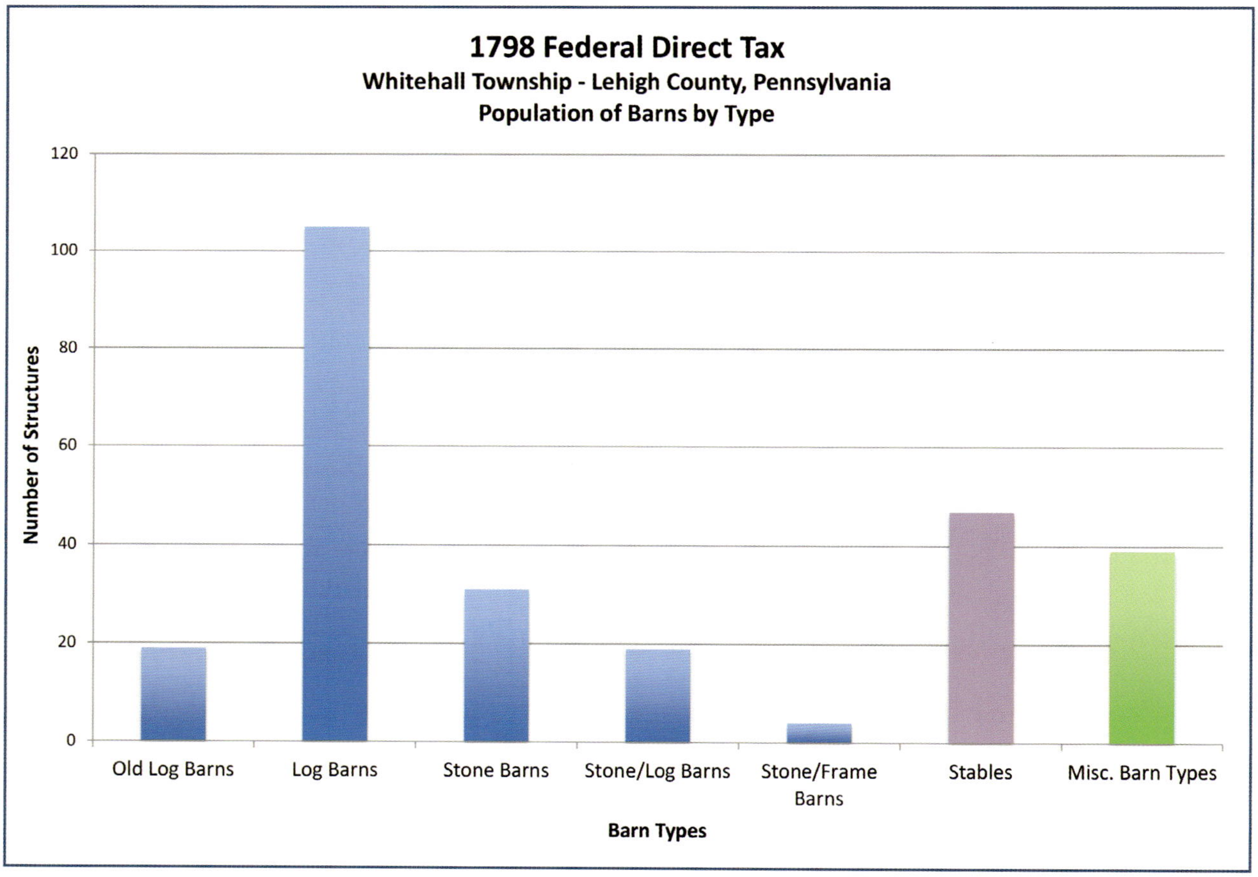
1798 Federal Direct Tax
Whitehall Township - Lehigh County, Pennsylvania
Population of Barns by Type
Number of Structures
0
20
40
60
80
100
120
Old Log Barns
Log Barns
Stone Barns
Stone/Log Barns
Stone/Frame Barns
Stables
Misc. Barn Types
Barn Types

The Rohrbach stone Sweitzer in Longswamp in Berks County was built just after the close of the eighteenth century or 1800 to 1810. This barn retains many of its original front wallboards with period wrought nails.

## Other Barn Surveys

I conducted three other surveys of areas surrounding Lehigh County. One survey of 106 barns was done in January 2012 in Longswamp Township, Berks County, to the immediate west of Lehigh County. Just one barn likely pre-dates 1810, or even 1800.

In January 2012 I examined an area near the southern border of Lehigh County —a strip of land a quarter-mile wide on either side of Route 100 from Hereford in Hereford Township south to Bally, a distance of about four miles. I found sixteen barns, none pre-dating 1810. The third test area was Lower Saucon Township in Northampton County that borders the east side of Lehigh County. Of fifty barns surveyed, just one barn seems to pre-date 1800—a rare, circa 1790 two-level log bank barn.

In the four surveys—in Lehigh County and adjoining areas—1,842 barns were counted. Just one definite barn from the pre-1800 era surfaced; one other barn was

The 1791 dated Caspar Maul stone Ground barn in the Oley Valley in Berks County is a classic example of the barn style. The roof structure is early German based. It has mow-stead wallboards secured with wrought nails.

possibly of that era, and four others of about 1800 vintage were also found. Any attempt to discern eighteenth-century barn- building traditions in the four survey areas is severely restricted due to the lack of early barns. In other words, it is far too late to make an in-depth study of barn-building customs of that century.

The areas at either side of Lehigh County, namely Northampton County and the eastern third of Berks County, have not been formally surveyed, but have been examined extensively for early barns, and almost no pre-1800 barns have surfaced. It is not until Oley Valley, especially Oley Township, in southeast Berks County, is encountered that a few 1780–1800 Sweitzers and Ground barns are found.

## Lancaster County as the Last Stronghold for Eighteenth-Century Barns

Lancaster County offers the greatest potential for surveying surviving eighteenth-century barns, especially Sweitzers. This is the earliest type of forebay barn that remains in the state, and this was the area where Sweitzers were probably first built. Many of these Sweitzers have the basic proportions, including steep roof pitches, of pre-1800 barns.

Sweitzers are also present in neighboring southern Lebanon County and nearby Berks County, offering a chance to learn more about the earliest surviving forebay barns in the state. They also offer the prospect of identifying regionalisms that will increase our understanding of early barn-building traditions.

Areas outside the hearth area, particularly York, Adams, and Franklin Counties, may also provide insights into pre-1800 barn customs. One example is a dated 1794 stone Sweitzer in Franklin County that includes a *liegender Stuhl* roof truss system. This tells us that the early roof format traveled far west of its place of origin in Lancaster County.

# Chapter Fourteen.
# The Spirit in Barns: Decorations and Mystery Marks

The Trexler stone Standard barn built circa 1820 in Longswamp, Berks County, has a distinctive mid-twentieth-century-style twelve-point barn star. Three such stars adorn the front forebay wall.

In my decades of studying barns, it became clear that these old buildings are more than mere aggregates of stone and timbers, boards and nails, and metal hinges. I sense a certain consciousness or spirit of the people who conceived and built them. One way that spirit is revealed is through decorations applied to both interior and exterior walls, such as white trim, half-circles above doors on forebay walls, and innumerable barn stars.

It was the Germans who most loved to decorate the material objects they created, whether furniture, metal-ware, treen-ware, or buildings. But they also added more occult-like symbols such as hex signs. What were artisans of long ago trying to tell us?

In *Hex Signs: Pennsylvania Dutch Barn Symbols and Their Meaning* (Yoder and Graves, 1989), the authors wrote that barn stars did not appear until the Civil War era, in the 1860s. In *An Age of Barns*, Eric Sloane

The stone Standard barn near Lenhartsville in Berks County has a six-point barn star with a date of 1819 in an architectural recess. Eric Claypoole repainted this barn star after the apparent original in 2000.

A stone Standard barn built about 1820 in Exeter Township in Berks County has an architectural recess with surviving vertical boards with a faded white border. No barn star ghost was found on the recess.

commented that barn stars, or what he called hex signs, "chust for pretty," started about 1830.

It turns out that neither Yoder and Graves nor Sloane were on the mark. Since those books were published, faint barn stars have been discovered on recessed wood circles just below the roof peak in barns from the early 1800s. These recesses were something of a revelation. In 2000, Eric Claypoole, a Berks County barn star master, worked on a stone Standard barn about one mile north of Lenhartsville in Berks County. One stone end wall had a faint date of 1819 on a recess just below the roof peak.

A great stone Standard barn circa 1810 in Douglass Township in Montgomery County retains its original two vertical boards with wide white border. Eric Claypoole climbed the wall and found no barn star ghost.

An apparent 1795 stone half transitional Sweitzer in the Oley Valley in Berks County has an architectural recess with no original boards. Many of the old forebay wallboards have no barn star ghosts.

In the fifteen-year period from 2000 to 2015, about fifty barns have been found with recesses, several of them dating from 1805 to the 1830s. Circles or ovals, they are often about 4 feet in diameter and recessed into the stone wall by 2–3 inches. A few of these barns have intact recessed boards with a faint barn star scribed onto them. And the fanciful scribe lines on boards on a barn with prominent recesses on both end walls, north of Kutztown in Berks County, can be readily seen. We assume that most, if not all, the recesses depicted stars in some form. Since the date marks cross-reference with other architectural features of the time, such as

A rare architectural recess found in early 2008 on a Standard barn in Washington Township, Lehigh County, has a plastered finish. It has an eight-point star and four four-point stars with a wide faded red border.

A stone Standard barn, found in 2004, in Longswamp in Berks County, has a unique condition that includes three circular recesses on the one end wall. None of the inset boards in this circa 1820 barn are original.

Principal Rafter Systems, we may assume that stars were applied on barns as early as 1805. These facts rewrite history, proving that stars were applied to barns throughout the nineteenth century.

One of the most interesting recesses, in Heidelberg Township in Lehigh County, is a sunburst within a 4-foot half-circle. It consists of a few original horizontal boards and a centered tab at the bottom with a faded date of 1823 and the Latin inscription I N R I (Jesus the Nazarene, King of the Jews) above it. Above that are five equally spaced sunrays interspersed with two tulips at the center and four rosettes, two at each side.

Another structure is a circa 1825 stone Standard barn off Longswamp Road in Longswamp Township, Berks County. The one end wall has three unique 4-foot-diameter recesses that lack their original boards. The center recess is just below the peak with the other two flanking it along the roofline.

The 1813 Joseph Peter Standard barn in Washington Township in Lehigh County has a rare original plaster recess with a wide red border and keystone forms at the cardinal points. The middle area has a centered, eight-point star and four four-point stars at the corners.

Another fascinating recess is in Pike Township, Berks County on a circa 1820 transitional stone Sweitzer.

A stone Standard barn in Heidelberg Township in Lehigh County, discovered by the author in 2007, has a very rare recess with a date on the bottom tab of 1823.

A Standard barn in Lower Milford in Lehigh County features a basement-level door with overhead hood that bears a date of 1858 with a regional teardrop barn star.

The recess is plastered with a wide border that has a blatant celestial basis on both the border and the center. The center has a seventeen-point star and initials J F at the middle. Beyond this out to the border are seventeen outward-pointing triangles and dozens of small circles; beyond that are dozens of thinner triangles where every other one is painted black. On the border are even thinner alternating black and white triangles.

Hundreds of Standard barns and some Sweitzers in the greater Lehigh and Saucon Valleys have hooded door entries at their house-side basement end walls. Many of the hoods include barn stars. Their dates range from 1840 to 1860. Many of the barn stars were painted over, but some door hoods retain their original appearance with the stars.

A unique plastered recess on a Pike Township barn in Berks County with a celestial motif features three concentric rings of points and a wide border with points. A ring of many connected small circles also appears.

An Albany Township barn has an original granary door adorned on both sides with several barn stars (whirling swastika type) and two dates, both 1857.

One two-section frame Standard barn in Longswamp Township, Berks County, has several barn stars including a 5-foot, twelve-point star on the front straw shed wall.

On the rear wall is a six-point star with "Pennsylvania hearts" between the points. In the middle is the name George Wetzel 1848. Two barns with stars on their exposed front walls have been found that have original dates right on the stars. One of these barns, in Albany Township in Berks County, has a star dated 1857.

A frame two-section Standard barn with front straw shed in Longswamp in Berks County has an unusual 5-foot diameter twelve-point star with each point intersected with a Pennsylvania heart, and a scalloped periphery.

A stone Standard barn found in 2012 in Albany Township in Berks County has three six-point stars with a rare date of 1857 on the middle star.

A post-1850 frame Standard barn in Lower Milford Township in Lehigh County is adorned with four large stars on its front wall with two very rare fourteen-point stars in the middle.

A wide-diameter eight-point barn star with wide pinwheel and eight crosses adorn an end wall on a frame Standard barn in Greenwich Township in Berks County.

The date seems to be authentic to the barn, as an original interior granary door, gaily decorated, was also painted with the date 1857.

Exterior barn stars vary considerably in size. A barn in northern Berks County north of Route 78, in Greenwich Township, contains a 6-foot-diameter, eight-point star with crosses between the points. Some barns have 5-foot stars, but 3-foot to 4-foot stars are the most common.

Stars with up to eighteen points have been seen, but this is rare. Three fourteen-point stars on barns have been found, including an excellent example of two stars on a small frame Standard barn, circa 1870, in Lower Milford Township in Lehigh County.

Although many barns have twelve-point stars, six- and eight-point stars are more common. They are seen in many different colors, including vibrant green. Some barn stars demonstrate regionalisms. For example, there are dozens of examples of a swirling form around the intersection of Lehigh and Berks counties. These stars typically have four large teardrops or petals emanating from the center.

Several Standard forebay barns contain barn stars that have been protected by the straw sheds built at the front. For example, the 1857 barn in Albany Township has a single barn star that was sheltered from the elements by a circa 1890 front shed. The star has about thirty years worth of erosion, suggesting that the front wall was fully exposed before the front shed was added.

Some of these protected barn stars were removed from their original front walls and recycled into other areas such as ceilings or flooring. This is another excellent means by which barn stars have been preserved.

The Albany Township barn has a front shed that has covered one of the three front wall stars for about 120 years. All stars have six points.

## Creators of Barn Stars

Just who were the makers of barn stars? The six-pointed variety goes back a millennium or two. Many examples can still be seen on European buildings dating to many hundreds of years ago. That the six-point barn star came to America in the seventeenth or eighteenth century is undeniable. It seems that the late-1760s Swamp Lutheran Church in New Hanover Township, Montgomery County, has one of the earliest depictions of an exterior six-point star in the state. A number of gravestones in the hearth also include these stars.

Barn stars were produced by both artisans and homesteaders. No professional nineteenth-century barn star maker has yet surfaced, but the well-known Milton Hill of Berks County painted barn stars in the 1905 to 1960 era. Harry Adams, who lived in Edenville in Berks County, worked for Hill for a while and says he painted barn stars on eight to ten barns. His first commission was for a barn about one mile up Mountain Road off Old 22. He used his own designs unless someone supplied a pattern.

At the front of the 1760s Lutheran Swamp Church in Montgomery County is a rosette as a part of the stone door surround at the front entrance. This is one of the earliest appearances of a rosette on any building in the state.

In 2010, Eric Claypoole plied his craft on a frame Standard barn near the Lehigh County border. He painted four large barn stars on the barn, including three on the front wall.

Barn star-like motifs can be seen on furniture such as this circa 1810 painted chest.

The Claypoole barn star is an eight-point star with pinwheel and eight small intersecting rosettes. Claypoole used five colors to make this star, which duplicated existing stars on the barn.

This large stone-to-peak Standard barn dated 1860 in southern Lehigh County is adorned with five large barn stars with five different motifs.

The premier barn star applicator today is Eric Claypoole, the son of Johnny Claypoole. Johnny painted stars on barns and other media for several decades. Eric has painted about one hundred barns, principally in Berks County and the greater Lehigh and Saucon Valleys. Eric is a talented artist and has helped to preserve a long-time tradition.

In his 2013 seminal book *Hex Signs: Myth and Meaning in Pennsylvania Dutch Barn Stars*, Patrick Donmoyer devotes an entire chapter to the "Claypoole legacy." His book is the most authoritative treatise on the subject to date.

## Barns Protected by Various Objects

Wallace Nutting may have first disseminated the idea that stars could be used to divert evil forces. He apparently heard from a single source, a man in the Bethlehem area, that barn stars were equated with "hexafuss." He then popularized this concept in his well-known 1924 book, *Pennsylvania Beautiful*. In his small pamphlet *Hex, No!*, Alfred Shoemaker stated that barn stars were in no way connected to warding off evil. This continues to be a topic of debate.

The 1840s stone-to-peak Standard barn in Upper Milford Township in Lehigh County has an unusual design on one mow-stead wall that consists of twenty or so interlocking circles.

The Shane barn in the Oley Valley in Berks County has its original mow-stead wall with a few rosettes with bold black and red areas.

## Barn Stars as Celebrations

Whatever their function, barn stars marked the pleasure a farm family took in the most important building in their lives. They must have understood that stars were connected to the sun and the sky itself, and that the success of their livelihood depended on the weather. Barn stars, perhaps more than anything, were a reminder of what life was about. The barn gave a farmer his life, and he celebrated his barn with various decorations, principally barn stars.

## Interior Decorations

The markings on barn interiors may be even more diverse, offering tremendous opportunities to study the handiwork of not only craftspeople but also the farmers themselves.

Marks are found everywhere, except on wagon floors and perhaps rafters. The most common areas are mow-stead walls and granaries. While there are too many kinds of markings to be categorized, the two most common decorative markings are six-pointed rosettes and grain tally marks.

Thirty to fifty percent of barns contain rosettes, which are thought to be purely decoration or to ward off evil, perhaps both. A barn in Heidelberg Township in Berks County, built around 1830, has a multitude of stars on its mow-stead walls.

The barn's current owner believed that the stars and other markings were strategically placed near wall openings to divert any "dark forces" loitering about. Others think such markings were "rehearsal spots" where artisans or farmers practiced their skills for later use on the exteriors. Other observers simply believe that farmers were doodling on the walls.

A mow-stead wall in a stone Sweitzer in Oley Township, Berks County, includes the rare presence of two rows of notches or marks. Every fifth mark is longer than the preceding four marks. Every twenty-fifth mark is longer than the fifth marks. They may have acted as tally marks.

## Tally Marks

Many farmers placed tally marks in or near granaries to keep count of the bushels of grain produced—a primitive accounting system. The marks were often arranged in rows, some of which are quite long. Pencils and what appears to have been chalk, or even knives, were used to make the marks.

An exceptional circa 1810 Sweitzer in the Oley Valley, Berks County, has dozens of vertical marks in two rows on a mow-stead wall. Every fifth mark was longer than the preceding four. Then every twenty-fifth mark was longer than every fifth mark. Two rows of such marks are seen on the wall, one on top of the other. The top row is 20 inches long while the lower row is 45 inches long.

Another exceptional system was apparently used in a circa 1870 frame Standard barn in Weisenberg Township in Lehigh County. Here a thin board, 12 inches high by 7 inches wide, was nailed to a vertical stud in the granary. The board consists of ten rows each of ten small holes. Small objects such as short pieces of straw were inserted in the holes. Perhaps each hole signified a single tally mark or perhaps a cluster of five. When the farmer was done with his work, it seems that the straw pieces were withdrawn.

A stone Standard barn built circa 1870 along Old 22 in Lehigh County has an apparent rare tally board. Ten columns and ten rows make one hundred holes for insertions of straw, an accounting system for farm crops.

## Fertility Marks

Many barns have so-called fertility marks or symbols. These are diamond shaped with sides about an inch long and some considerably larger. Some marks have dots in the middle. Others consist of two or three concentric marks. Sometimes a few marks are clustered together. These symbols appear on wood elements such as doors in basements of bank barns or on upper floors. Occasionally they are found on timbers of covered bridges.

Fertility marks have ancient connections. The most prominent fertility mark in the hearth is actually a stone on the end wall of a circa 1820 stone-to-peak Standard barn near Maple Grove in Longswamp Township, Berks County. The diamond-shaped stone projects a few inches beyond the regular wall surface. Each side of the stone is perhaps 12 to 15 inches long. These fertility marks signalled a farmer's wish for high crop yields and milk production.

A Standard barn in Williams Township, Northampton County, has a stable wall painted with galloping horses. Three live horses at the doorways make sure everything is running according to plan.

On an open-frame carriage shed of the mid-1800s near Blue Ball in Lancaster County is a rare two-by-two diamond within a larger diamond fertility symbol on an end wall. Such signs reach back to antiquity.

A unique stone fertility symbol appears on a circa 1830 stone Standard barn in Longswamp Township in Berks County just below the peak on the end wall. The farmer wanted his farm to have a high level of fecundity.

In a circa 1880 frame Standard barn on the basement level in Lower Saucon Township, Northampton County, are several triangular stones of various sizes. Did the three sides to each stone indicate the Trinity?

The superb Wagner barn near Hellertown in Northampton County also has a diamond-shaped fertility stone on its rear wall. A nearby circa 1890 frame Standard barn has many exterior facing stones with peculiar shapes on three foundation walls, obviously intended to make a statement.

# Chapter Fifteen.
# Preservation

Barn people feel deeply when one of their beloved farm structures disappears from the landscape. Barns form a deep part of their awareness of the past and are part of their cultural inheritance. Barns, they often think, are integral parts of rural landscapes. Without them, something important is missing, and their disappearance can leave us feeling bereft. How can this be stopped?

It can't. Not enough people care; they take barns for granted. If we care about barns, we care about more than just the sum total of their physical components. Barns occupy spaces where certain things happened. Can we imagine the conversations that took place when farmers and their spouses selected places to live and the exact choices they made about where and how to build their barns? Barns are a record of the concrete decisions these early people made. Will we honor that?

Select trees were felled. Later, timbers were floated down rivers and creeks. Someone took an axe out and transformed the trees into beams in specific ways, and later made tenons and mortises so the timbers could be joined in "holy wedlock." Can we picture the multitude of actions that led to the construction of timber bents? Can we envision other things such as basement walls being laid and stable wall doors being built, and all the staked mangers and mangers' troughs being put into their correct places? Can we conceive of mow-stead walls and exterior walls and roof lath or roof boards being set down? Rafters needed to be positioned. Can we even imagine the ceremonial tree in the "topping out" ceremony being joined to the top of a rafter pair after all the barn pieces were put into their final resting spots? Will we honor that?

Time is the greatest enemy of barns. Those of us who care about the fate of barns are racing against time. Just since 1975, when I started to observe barns, many hundreds have disappeared in New England, New York, New Jersey, and Pennsylvania, especially in the hearth area. Time will continue in its forward march, but we can make a conscious decision to record these structures. This is the single most important thing we can do to preserve barns. But surely, some may ask, it costs thousands of dollars to preserve a barn—to replace a leaky roof or rotting foundation. That's not always the answer.

In the spring of 2014, Phil Schaff, a resident of Franklin County, showed me a truly magnificent brick ender bank barn in that county. The one brick end wall with its spectacular patterning was the single greatest example of its kind I have ever seen. Schaff, who has photographed more than a hundred barns in the county, emailed me to say that the barn had burned to the ground in the early summer of 2015. If the barn owner had spent $50,000 to $100,000 to renovate the barn before it burned down, but had not documented the original barn, what would we be left with? Spending a great deal of money on a barn is not always the best move, not by a long shot.

An even better idea is to document the rarest and the earliest barns, as these examples help us understand how barns evolved. But how would we identify these barns? The answer is that extensive surveys are required, as they reveal the most important barns in any community. It takes time and money, and a lot of both.

Most of us can learn how to record a barn. The Historic Barn and Farm Foundation of Pennsylvania helps people in this regard. A number of barn tour guide booklets show how individual barns are recorded. In addition, a section of its website shows people how to fill out documentation forms (www.pahistoricbarns.org).

A frame Ground barn circa 1890 in Upper Macungie Township, Lehigh County, had been abandoned when it was purchased by the current owner in 1975. It was soon converted to a beautiful home. The addition on the left was built from recycled timbers.

## Adaptive Reuse

Adaptive reuse is another method of preserving barns. Barns of many different types and sizes can be repurposed as residences, apartment buildings, office buildings, places for wedding receptions, and even riding stables. Many barns offer expansive spaces that may be converted into any number of uses, and reveal their structure in an exciting way.

In the past few decades, hundreds and perhaps thousands of barns have been converted into residences. Eric Sloane had a barn transferred from Canada to his place in western Connecticut that he used as an artist's studio. The economist Stuart Chase, another New Englander, lived in a converted barn in Redding, Connecticut. Artists and authors are often drawn to barns.

One noteworthy example took place in the mid-1970s in Upper Macungie Township, Lehigh County. A contractor and long-time resident of Kuhnsville, John Kuhns and his wife, Carol, converted an unusual circa 1890 gambrel roof frame Ground barn into a spacious residence. They added a frame section and have made this their home for forty years. John, a lover of old barns, worked on barns in the greater Macungie area for many years.

Kevin Durheim, of Lower Macungie, converted a rare area 1870s brick Standard barn into a home. In still another example, in the late 1990s a local contractor dismantled a two-level bank barn in Heidelberg Township, Lehigh County, and moved it to a site along Route 378 in Bethlehem, Northampton County. A new building was erected using many of the timbers from the disassembled barn. Elements of the bents and other features were salvaged and became the home of the Lafayette-Ambassador Bank. Although the structure is now only one story, it can still be called a "bank" barn.

In adaptive reuse, most of the basic structure is saved, and thus the information is preserved. This is significant, as it becomes part of the historical record. The previous example notwithstanding, it is best for barns to stay in their original locations. That's because a barn's location sheds light on the circumstances under which it was built. When you move a barn from its original location, much cultural information is lost.

## Maintenance

Barns are in various states of repair. Those in excellent condition are usually being actively used for storage. But even those that stand empty will maintain their structural integrity if they simply have a good roof to keep out rain and snow.

The following is one protocol, among several others, for maintaining a vernacular barn. The closer this outline is followed, the better "structural health" a barn will have.

Someone buys an old homestead, complete with a historic house, a historic barn, and perhaps a few outbuildings. The new owner may have bought the place in "as is" condition. In the first year or two of ownership, the barn should be carefully examined. This includes all exterior aspects including the roof, stone pointing, or vertical siding if the barn is frame. Inside, examine the structural elements including the roof, bents, and longitudinally oriented timbers if it's a frame building. All timber joints should be tight with sound wood throughout the beam connections. The basement also has structural parts such as summer beams and joists. Joists are especially vulnerable to decay where they enter the rear stone basement walls.

Ideally, if a barn owner can afford to do so, he or she should hire a structural engineer familiar with historic buildings to carefully examine all structural aspects. The barn owner may do all necessary repair work at once, or piecemeal work over a few years, but the most important things should be taken care of first.

No matter what the state of repair, the owner should perform regular maintenance checks. Time exacts its toll on any building, and bad weather can affect barns in deleterious ways. After a storm, the barn owner should check for possible structural disruptions outside or inside. Importantly, the owner should periodically check the barn floor for evidence of roof leaks. Vigilance is the order of the day. Never forget the old saw: a stitch in time saves nine.

The other side of the coin involves the 25-foot rule. Remove objects that hold moisture (including snow) on the exterior, including those within 25 feet of the barn, such as tree branches. Branches thrown about in a storm can wreck havoc on the roofs and walls. And it's easy to let things accumulate near or against the outside walls, but these objects leak water into the barn wood and invite bug infestation.

Another canon of good barn maintenance is air circulation. Doors—especially the big rear-wall wagon doors and the winnowing doors on the front walls of bank barns for cross ventilation—should be periodically left open in good weather to keep interior moisture in check. Admittedly, not all barn owners can perform this chore routinely, but the more often it is done, the better.

Simple, regular maintenance will extend the life of a historic barn, and in turn, the amount of information we have at our disposal. When it comes to old barns, time is not on our side. It never has been.

# Epilogue

In the end, the barn builders' thoughts fell away. They cleaned up after themselves—all the loose and discarded timbers, bits and pieces of boards either unwanted or forgotten, chunks of stone, some big, some small, that never found their way into a wall. Unused nails, too, were gathered. Uneven mounds of earth were smoothed once again. They looked up and saw that all they had executed went according to plan. They knew that only oak could reflect the splendor and distinction of their building because it would endure over the coming decades and centuries. A final walk to the farmer, and settlement was made. The clasp of hands gave them each license to move on.

And then the new barn was filled—animals in the basement; farm crops, many in number and type, in the upper loft reaches; threshing was done on floors between mows; and subsequent hearth homesteaders repeated the process.

The barns changed again and again. Those built during the first hundred years of occupation disappeared and will never be exactly known. Their sizes and lengths, the twists and turns of the timbers and how they met their neighbor beams, are lost forever. Deliberately placed stones met the same fate.

Some later barns survived with a precarious hold on the land. Animals and farm crops still had their places, but not the same places. Ground barns gave way to their larger cousins, the bank barns. The hearth was expanding and soon Sweitzers and English Lake barns swarmed the land. The forebay idea had staying power, and Sweitzers later yielded to Standard barns. These later forebay types dominated landscapes in the hearth for more than a century.

On untold numbers of barns, especially in the eastern hearth, the broad surfaces of forebay walls begged for adornment, leading to painted decorations such as barn stars in a variety of expressions. Carvings and other marks appeared on mow-stead walls and granaries. Barns became quiveringly alive.

Near clones of the basic Standard barn broke out in the hearth in the post-1820 era and spelled the golden age of agriculture. It was barn-building exuberance to its fullest extent, resulting in an exceptional range of types and sub-types. Not until the first quarter of the last century did the Standard barn surrender to more modern designs.

Like manifold fingerprints on the hands of the populace, regionalisms left traces of vernacular styles all around. Barns in some areas had stone walls, while others were frame. Details evolved, such as the arched longitudinal beams in southern Bucks County barns. All these variations added personality to the landscapes.

Careful study of construction features and dendrodating science allowed barns to be dated accurately. They supply answers to the important question, "How old is it?" Fifty years ago, all we could do was wonder. That has all changed.

Carl Jung knew it is not what we see, but what we see in something, that gives meaning to an object and defines what we appreciate, respect, and know. And it was Eric Sloane who advised us "to see where others only look" (Sloane 1954). May this book help people see the Pennsylvania hearth's barns with new eyes. Their forms and timbers contain messages from a forgotten past and can be read by anyone who takes the time.

# Appendix

## Curious about the History of Your Old Homestead Barn and House?

Contact Greg Huber for a historic building evaluation or to prepare a National Register nomination. Huber does dendrochronology examinations and also compiles reports on architectural barn and house histories for owners of old homesteads in all areas of the Eastern United States. Reports include construction type, history, construction date, and other historical aspects.

For more information, visit www.past-perspectives.com and www.easternbarns.com. Readers who have questions or musings about any of the content in this book may contact the author at greg@easternbarns.com, or send queries to P.O. Box 82, Macungie, PA 18062.

## Sign up for the New Barn News and Events Newsletter

In August 2017, Eastern Barn Consultants started a new newsletter: *Barn News and Events Newsletter*. The newsletter is emailed to subscribers every two months and contains three or four interesting articles about unusual barn stories and visits that Eastern Barn Consultants make to various geographical areas—both near at hand in Pennsylvania and other states, and even Canada. This is information that you will not find elsewhere. Occasionally, articles will tell stories of special barn preservation efforts by concerned barn people and certain groups. As of the November 2018 newsletter nine issues have been sent out to many hundreds of subscribers in several states, as well as Canada and Australia.

The newsletter also has announcements of barn tours and publicly presented lectures during the course of the year. In addition, readers are kept informed as to what new articles have been recently written and published by certain organizations dedicated to historic preservation.

To subscribe—both individuals and organizations alike—email us at greg@easternbarns.com and tell us you would like to receive all future issues of the newsletter.

# Glossary

**Basement.** This floor level in two-level or bank barns occupies the entire area below the second floor and was used to stable animals.

**Bay.** The space between adjacent framing units or bents. For example, there are three bays in a frame barn with four bents. In stone barns the end bay is the area between the last bent and the adjacent stone gable wall. Thus, two end bays appear.

**Bent.** A major transverse framing unit on the second floor of bank barns that consists of partial-barn-width girts, vertical side wall posts, mid-point posts, braces of varying lengths and, most often, built-in ladders. They were assembled on the barn floor or ground and then raised with pike poles. Ground barns also have bents.

**Brace.** These diagonal members are either relatively short or quite long. When they are short, these beams connect girts to either the wall posts or the other posts of the bents. They prevent the bents from lateral racking. Braces also appear in other areas such as the roof plane, where they are called wind braces. They can also connect purlin plates to queen posts. When braces in bents are long, they span from the floor sill to the upper girts. These braces are often considered to be Germanic in nature. In Pennsylvania barns, braces are usually pegged at each end.

**Built-in ladder.** Original elements usually centered on the bents adjacent to the wagon bays. Thus, Pennsylvania barns often have at least two built-in ladders; three ladders frequently appear in four-bay barns. Ladders often have oak rungs or cut-out areas in horizontal boards that function as rungs. Occasionally the ladders extend to the roof. Quite often they extend to the straining beam where vertical queen posts appear.

**Collar beam.** This horizontal beam unites rafters in a rafter pair. Collar beams are almost always associated with vertical queen posts in German related barns, also called *stehender Dachstuhl*. Often these collars were removed decades ago when hay tracks were installed in barns in the last third of the nineteenth century or later. Collars were tied to rafters via lapped, half-dovetail joinery or mortise and tenon joinery.

**Common Rafter System.** Major structural, basically same-sized roof members, often oak. The rafters virtually always come in pairs and in almost all pre-Civil War barns and fairly often in post-Civil War barns, rafters are pegged at the roof peak. Rafters are joined to the wall plates.

**English Lake District (ELD) barn.** This is a very rare barn type in the greater Lehigh and Saucon Valleys. ELD barns are much more prevalent in the more southern English counties—southern Bucks and Montgomery, Chester, and Delaware Counties. These banked, two-level structures have a basement and an upper floor without any forebay. The upper level is typically stone, and rarely frame. The greatest difference between these barns and Pennsylvania barns is that the front walls in most English-type barns are all stone on the front wall—from the ground to the top of the wall. In addition, a short roof projection is seen at the top of the stone basement wall at the front, called a pent roof. It extends across nearly the entire length of the barn. The pent roof protects the stable wall doors.

Extended forebay barn. A Pennsylvania barn that has been enlarged by amending or extending the barn beyond the basic Sweitzer and Standard framing limits. One example is an original condition Up-country Posted barn in Lower Saucon Township. Six types of barns comprise the Extended class.

Forebay. The area of a second floor in a bank two-level barn that cantilevers over the stable wall.

Gable. The wall at the end of the barn that extends to the roof peak. Gable walls, also called end walls, are perpendicular to side or eave walls. Thus, four-sided barns have two gable walls and two side walls.

Granary. Often located in the forebay of a Pennsylvania Sweitzer barn, and in Standard barns often in both in the forebay and a few feet to the rear of the "designated" forebay area. One granary wall is continuous with the mow-stead wall that flanks the threshing floor. The front wall of the granary is actually part of the front wall of the barn. Granaries contain individual compartments to store different kinds of grain. Ceilings are about 6½ to 7 feet high. Farm crops were stored above the granary room.

Ground barn. A Ground barn is virtually always a one-level barn. There are no regular full-height floor levels below the wagon floor. Ground barns were likely the earliest form of barn in Pennsylvania. The earliest surviving barn of the type is the superb Heckler barn dated 1761. The barns were constructed until late into the nineteenth century. Construction materials were log, stone, frame, stone and log, log and frame, and occasionally brick. Three-bay construction was the most common but several two-bay barns are also known. Likely 300 or more persist on the landscape (210 Ground barns have been identified).

Joists. Transverse ceiling timbers in basements that extend from the rear wall over the stable wall and out to the front forebay wall. Usually joists are supported by summer beams. Joists are usually oak and can be 10 to 18 inches wide. They are often left in the round at their sides, but are hewn square at the front of stable walls, apparently for aesthetic reasons.

*Liegender Dachstuhl.* Early German style roof truss support appearing in early barns built from 1750 to 1790, and seen in some Standard barns as late as 1850. The defining timber is the tapered, truncated Principal rafter. Their tops are notched to receive the purlin plates. These rafters almost always are associated with common rafters that appear above the principals. Another timber of these distinctive trusses is a straining beam that joins the tops of opposing principals in a pair. At the sides of the tops of the principals are slots into which the straining beams are placed. End braces also appear.

Longitudinal. A barn timber placed parallel to the ridgeline. Full-barn-length longitudinal timbers are purlin plates, wall plates, summer beams, or front forebay wall sills. Other longitudinal timbers that do not run the full barn length are girts or horizontal timbers at the front wall in Standard barns. Longitudinal timbers are placed at ninety degrees to transverse timbers.

Mow-stead wall. These walls are part of the bents that flank wagon bays and are most often 5 to 6 feet high. They consist of horizontal pine boards in many post-1820 barns, or oak boards in certain pre-1820 barns and are secured by wrought, transitional, or cut type nails.

*Oberdenn.* The area over the *denn* or threshing floor in a German barn. A series of *oberdenn* beams were placed longitudinally 11 to 12 feet over the wagon floor. The beams stretch from the tops of the upper partial-barn-width girts of one bent to the upper partial-barn-width girts of the opposite bents. Above these beams were placed wide boards, above which farm crops were stored.

*Oberdenn* longitudinal timbers. Timbers in barns with early German roof supports where beams, usually in line with the ridge, stretch over the various bays 10 to 11 feet above the floor. The beams are substantial, averaging 10 by 10 inches. No Standard barn has ever been identified with this type of timber.

*Peiler eck.* The German word for pier corner. These are ell-shaped stone formations at the front corners of stone and frame forebay bank barns; they give extra support to the gable walls.

Pennsylvania forebay barn. A two-level banked structure with a cantilevered forebay or projection over the basement-level stable wall. The three categories of Pennsylvania forebay barns include Sweitzer, Standard, and Extended Forebay.

Principal Rafter System. Large rafters aligned with bents and in end bays, seen in English or German-based barns built from about 1790 to 1840 and likely into the 1850s in many Saucon Valley barns. Barns with such roof structures have two to five Principal rafter pairs, each composed of two large rafters. Trusses are either in line with the framing units or bents or they appear about mid-way in end bays. They alternate with considerably smaller common rafters.

Purlin plate. This is a barn-length or longitudinal timber that extends from gable wall to gable wall. They usually come in pairs, one per roof slope. Purlin plates support rafters and are often spliced. They are often sectioned and are joined by refinements called scarf joints. Barns featuring Principal Rafter Systems have purlin plates that are decidedly different from those in barns with Common Rafter Systems, in that they are often sectioned but are joined to the sides of the Principals rather than spliced.

Queen post. A vertical or angled timber that extends from an upper tie beam of a bent or structural framing unit to the soffit or undersurface of the purlin plate. Thus, the queen post supports the purlin plate. Canted queen posts were often used after 1840, but appear before that time (see *stehender Dachstuhl*).

Side wall. The side or eave wall is the wall that appears between the gable walls. In the great majority of Pennsylvania barns the main wagon doors appear on the rear side wall. The front forebay wall should be considered a side wall.

Sill. A piece of timber at the edge of the floor. Individual sills acting in concert with each other constitute a sill system. They appear along the edges of threshing bays and wagon bays. They also appear at the bottom of the floor at the front wall of the forebay.

*Spriggel* bar. This is a horizontal, square piece of wood located to the side of a horse stable door. About 3 inches on each side and usually made of oak, the bar sits in a channel in the stone wall, usually surrounded by a wood box along which the bar slides. The bar may have been placed at the height where the two door halves of a stable wall door meet to prevent horses from gnawing the tasty pine wood. *Spriggel* bars are more commonly found in barns built before about 1850 or 1860.

Stable wall. The stable wall or *stahlwand* at the basement level of forebay Standard barns is set back from the front edge of the end wall by 4 to 10 feet. It contains several animal doors and one or two feed-entry doors. Often these doors are halved, with a top and bottom. Sweitzers also have stable walls.

Standard barn. Characterized by a symmetrical roofline, Standard barns were first constructed in the 1790s and were far more commonly erected after about 1820, until the late nineteenth century and somewhat beyond. The forebay width is often about 4 feet but can measure 6 to 8 feet or more, depending on the specific barn and barn type. The transverse framing units or bents are contained within the front forebay area. Ten types of barns comprise the full Standard class.

*Stehender Dachstuhl.* An early German vertical queen post seen in pre-1830 built barns. These posts are usually quite thick and are placed vertically, emanating from the upper tie beam of a timber frame bent. They are virtually always associated with horizontal collar beams. Sometimes the posts are flared or gun-stocked (for a rather dramatic visual effect) at the top to receive both the purlin plate and the collar beam.

**Straining beam or queen post tie beam.** These transverse beams stretch from one queen post to another above the bent. They were often included in barns before about 1860. In many barns these ties were removed when metal hay tracks at the roof peak were installed after the Civil War. These tie beams never, or very rarely, appear between canted queen posts. They also appear in Germanic roof structures.

**Summer beam.** The barn-length longitudinal ceiling beam, also called a girder, in the basement that forms the main support of the second or loft floor. Actually, joists support the floor above, but the summer beam supports the joists that appear above the summer. They are often oak and spliced in one or two spots.

**Sweitzer barn.** Sweitzer barns may have been erected before about 1750 but were commonly built in the 1780 to 1810 era. In certain areas such as the Mahantango Valley north of Harrisburg, Sweitzers were erected into the 1830s and 1840s. Sweitzers have asymmetrical roofs where the distinctive appendage-like forebay at the front creates the roof asymmetry. Sweitzers are common in much of Lancaster County and also Lebanon County and western Berks County. They become increasingly rare out to the Delaware River. Four types of barns comprise the Sweitzer class.

**Transverse.** Timber placed perpendicular to the roof ridge, or side wall to side wall. Full-barn-width transverse timbers are upper tie beams and basement ceiling joists. Another type of transverse timber that does not run the full barn width is a partial barn-width girt that appears in bents below the upper tie beam.

**Upper partial barn-width girt.** Sometimes called scaffolding beams, they appear in bents of Pennsylvania forebay barns and are seen a few feet below the upper tie; they stretch between a side wall post and a post positioned somewhere between the end posts.

**Upper tie beam.** This is the transverse beam at the top of a bent that spans the entire width of the barn and appears at the top of the bent, or more rarely is dropped 1 to 2 feet below the tops of the posts. It is almost always in line with the wall plate or just above the plate. Upper tie beams are usually associated with early English barns in New England, and they tie the longitudinal side wall bents together. In Pennsylvania these beams rest on side wall posts of frame barns and may or may not rest on posts in stone barns. They are placed into position after the bent is raised to its vertical position. This type of beam is found in Pennsylvania forebay barns built before the Civil War, and occasionally beyond.

**Wagon doors.** Wagon doors were built in three principal ways – harr hung, frame and board and batten types. See chapter ten (Construction Elements) for further reading on these important parts of barns.

**Wall post.** These major vertical timbers are part of the front walls in stone Standard barns and on front, rear, and gable walls in frame Standard barns. Posts usually appear in bents at both the front and rear side walls.

# Literature on Barns in Pennsylvania

At the start of the second half of the twentieth century, there was virtually no literature on vernacular North American barns. It seems no one was interested in these largely forgotten buildings.

First, to set the record straight, the word "barn" has its own history. According to Don Perkins, the author of *The Barns of Maine*, the word is an old English term that harks back to medieval times or the 1100 to 1400 AD era. He says that barn has "its origins in grain." Barley and barn are quite similar in spelling. Further, " . . . barn is from the combination of the old English expressions for barley (bere) and house or store (aern). So combining them results in 'barley house' or 'barley store'" (Perkins 2012). Thus, 600 to 900 years ago, livestock did not actually have a place in a building properly called a barn.

By 1950, even in Europe, monographs on barns were close to non-existent. One exception is a work by Reverend Robert Willis (1800–1875), *A Description of the Sextry Barn at Ely, Lately Demolished.* Another was a treatise entitled "Medieval or 'Tithe' Barns" by Francis B. Andrews dated March 21, 1900. This was published by the Birmingham and Midland Institute of the Birmingham Archaeological Society in the publication *Transactions, Excursions and Report*. Many of the barns discussed date prior to 1500 AD. Willis states that barns at places such as Cholsey, St Leonard's, Ely, and a few others were some of the most important structures. Most of the text is descriptive and few of the barn dynamics that underlie their appearance were considered—or even known at the time.

The following resources, listed chronologically by year, describe barns in Pennsylvania, mostly forebay-style structures. This comprehensive list offers a wide-ranging view of the earliest articles and book-length research published on barns in the state.

## Frederick Watts, First Article

Pennsylvania was one of the major settlement areas in North America and has been the home to barn building traditions since the middle third of the seventeenth century. Despite its prominent position in being a cultural hearth and distinctive building construction area, the barns in Pennsylvania had never been the focus of any study until 1855. It was Frederick Watts (1801–1889), a Carlisle, Pennsylvania, agricultural reformer and lawyer who, with deep connections to things agricultural, wrote a treatise on the barn. In 1840, with the help of Cyrus McCormick, he demonstrated the operation of McCormick's reaper for the first time in Pennsylvania. Watts was elected the first president of the Pennsylvania Agricultural Society in 1851. He was the US Department of Agriculture Commissioner of Agriculture from 1871 to 1877 under Ulysses S. Grant. His long article on the bank barn appeared in the *York County Star* and *Wrightsville Advertiser*. About the year 1860 he designed and had a barn built. A photo of the barn of that year survives and it is one of the earliest photographic images of a barn in Pennsylvania. He also wrote a ten-page treatise, "The Pennsylvania Barn," in the 1864 US Department of Agriculture Report.

## Marion Learned, Article on German Barns

Nearly sixty years later, in 1915, the first academic, Marion Dexter Learned, wrote the next article on Pennsylvania barns. He was born on July 10, 1857, near Dover, Delaware. He had an English and Welsh colonial background. Learned was a member of the Dickinson College class of 1880, in Carlisle, Pennsylvania. After graduation, he served four years as language instructor at the Dickinson Seminary in Williamsport, Pennsylvania. In 1887 he was awarded the first doctorate degree in German bestowed at Johns Hopkins. In 1895, he achieved status as professor of Germanic languages and literature at the University of Pennsylvania. He served as the pioneering chair of his department until his death in 1917. Learned's life work involved the study of German-American cultural contributions. This was his connection to barns that led to his twelve-page article in the University of Pennsylvania Lectures, 1913–1914: "The German Barn in America." He said that the dominant type of the Swiss barn was the so-called bank barn.

## Wallace Nutting, *Pennsylvania Beautiful*

Wallace Nutting's 1924 book, *Pennsylvania Beautiful,* describes barn folklore and includes a discussion of the origin of barns stars (hex stars). Nutting is credited as the first writer in a well-distributed book to use the word hexafoos. The term "hex sign" that apparently came from this word was based in the belief that hex signs on barns could ward off evil.

## Thomas Jefferson Wertenbaker, *The Founding of American Civilization: The Middle Colonies*

Thomas Jefferson Wertenbaker (1879–1966) was a prominent historian and a professor at Princeton University for thirty-seven years. In his well-known 1938 book, *The Founding of American Civilization: The Middle Colonies*, Wertenbaker proposed some general ideas on the roots of the Pennsylvania barn. He claimed that there were direct connections to Germany and Switzerland and thought certain building types, such as the Upper Bavarian house, had a particular propinquity to the Pennsylvania barn. The residence, barn, and stable were one building unit. But, as Robert Ensminger said of Wertenbaker's statements, "the barn (as) part of the highland house had been retained almost unchanged, in America, which makes his statement that the forebay may have been an afterthought confusing and inconsistent with his general logic" (Ensminger 1992). It is clear that the source of Wertenbaker's confusion lies in the fact that he did not commit himself to a proper level of European field research so that more definitive statements could be offered about the genesis of the forebay barn.

## Eric Sloane, *American Barns and Covered Bridges*

It was not until 1954, when the famed painter of Americana, Eric Sloane, wrote *American Barns and Covered Bridges*, that barns as architecture received any press in a single book. Separate chapters were devoted to the wood used and the tools that helped define the appearances of these buildings. A thirty-page chapter included illustrations of barn types, including the "continuous architecture" of New England, the Yankee version, tobacco barns, round barns, and also the Pennsylvania "Standard." One page depicts two sketches of Pennsylvania stone bank forebay barns. In his signature way, Sloane said that these barns are "banked a hill, stone to weather."

## Alfred Shoemaker, *The Pennsylvania Barn*

One of the foremost scholars of Pennsylvania German-American culture, Alfred Shoemaker was the editor of *The Pennsylvania Barn*, the first book exclusively devoted to barns in Pennsylvania. One of four authors, Shoemaker contributed twelve short articles that appeared in 1955. While offering a broad overview of barns in Pennsylvania for the first time, it should not be considered scholarly

John Heyl, here in his home in Booth Bay Harbor, Maine, was a historic preservation architect in the Lehigh River Valley. He lived just six weeks short of his 105th birthday.

in nature. But it set in motion a movement to appreciate barns as a historic cultural resource.

Possibly the most important chapter in the book was the contribution by historian Don Yoder, "Through the Traveler's Eye." Yoder cites observations made by both eighteenth and nineteenth century individuals. He was one of the driving forces behind the establishment of the Kutztown Folk Festival in the early 1950s.

## Charles H. Dornbusch and John K. Heyl, *Pennsylvania German Barns*

In 1958, *Pennsylvania German Barns* appeared on the literary landscape. Although the copyright date is 1958, the book was actually the 1956 annual volume Number Twenty-one of the Pennsylvania German Folklore Society. Co-author Charles Dornbusch was a young Chicago architect in 1940 when he was commissioned to take photographs of barns in several counties of southeastern Pennsylvania. Several years later, the German Folklore Society looked at his photos that were part of the collection of the American Institute of Architects. The decision was made to contact John Heyl, who was the recipient of the A.I.A. Medal Award at Harvard in 1933, to supplement the Dornbusch material that would result in a book.

Heyl went assiduously about in the early to mid-1950s taking a number of photos of barns in the hearth area. From this, together with the Dornbusch barn images, hundreds of photos had been amassed. Heyl assigned himself the difficult task of classifying the barns in the collection. No scholar or observer of North American barns had ever attempted any codification of barns in a broad geographic area. He almost completely relied on his own experience and powers of observation. He had the advantage that a number of individual barns included dates in their construction. Working from that base of information, he was the first to give us any sense that certain barns were built in particular periods.

The classes of barns and their names that Heyl created generally aligned with later research on barns in Pennsylvania. One class of barns, however, that was assigned incorrectly was Type E, a stone masonry two-level barn. This class is now known as the English Lake District barn, as they originated in England's Lake District. These barns do not have forebays, but some of their features match aspects of stone forebay barns. To this day, although hundreds of these barns can be found in mostly English based areas in certain counties in and around Philadelphia, no scholarly treatise has ever been done on this distinctive barn type. They were built at least as early as the 1760s.

Without question, Heyl's efforts were trailblazing and formed the basis for the work of Ensminger and others in sorting out and refining basic barn types.

## Henry Glassie, Articles on Pennsylvania Barns

Beginning in 1965, Henry Glassie, premier folklorist and a former professor at the University of Pennsylvania and the Indiana University Bloomington, wrote two articles on barns in Pennsylvania: "The Pennsylvania Barn in the South" and "The Double Crib Barn in South-Central Pennsylvania." Glassie proposed an evolutionary sequence of the development of the forebay barn. He believed that the initial form in the state was the one-level Ground barn, followed by Ground barns with an overshoot, and then a two-level barn with an unsupported forebay. He stated that the front overhang on these barns resembled certain structures in European countries including Switzerland and England. He concluded that the architectural elements that comprise the fully developed forebay barn were an American cultural invention. This refutes Ensminger's findings, where he identified hundreds of forebay barns in eastern Switzerland.

Eric Sloane was the first in North America to author a book on barns. In 1954, his book *American Barns and Covered Bridges* put these quaint old structures on the proverbial map. (Courtesy of the Thacker family)

## Eric Sloane, *An Age of Barns*

Eric's Sloane's classic 1966 book, *An Age of Barns*, illustrated for the first time, by means of strong pen and ink sketches, a broad array of North American barn types. Three major barn types were included: the side wall English Ground barn, the Holland Dutch three-aisle barn, and the Pennsylvania forebay two-level barn. This book may rightfully be called a celebration of the many North American barn types and accessory buildings that have graced the cultural landscapes of America. It includes many details of structural components and also floor plans.

One of his nine brilliant color oil paintings is a view of what Sloane named a Pennsylvania bank barn. A stone English Lake District barn (not named as such) and a few log barns with forebays are also depicted.

The Elmendorph Dutch-American barn, built in 1851 in Ulster County, New York, is a late example of an early New York barn-building tradition.

### John Fitchen, *The New World Dutch Barn*

John Fitchen was a professor of fine arts at Colgate University and a New York architect. His book, published in 1968, was the first in-depth technical study of a major barn type in North America. He briefly cited the works of Shoemaker and Dornbusch–Heyl.

### Joseph Glass, *The Pennsylvania Culture Region: A View from the Barn*

In 1969, Joseph Glass completed a dissertation as a partial requirement for a Pennsylvania State University PhD, which he received in 1971. His research culminated in his book, *The Pennsylvania Culture Region: A View from the Barn* published in 1986. He used the Pennsylvania barn region as a surrogate measure of the apparent extent of the Pennsylvania culture region. In doing so, a geographic area of about 20,000 square miles was covered. It included virtually all the southeast quadrant of the state, a small area in west-central New Jersey, most of northern Maryland, and a wide tongue of land extending along the northwest boundary of Virginia.

Glass generated information on barns at 530 homesteads due to fifty-three grid intersections that he instituted as sampling points to create a database. From this, he created eighteen maps where each map addressed itself to certain barn characteristics—the first map of its kind. His maps show what he called the domain of the Pennsylvania culture area. This entailed geographic sections in the Mid-Atlantic region under the influence of German cultural manners of folklife that prominently included barns. This was one of the most significant findings Glass made in his book.

Glass did not make a distinction between Sweitzer forebay barns and Standard barns. He only recognized two forebay styles—the open unsupported forebay type and the fully supported (at both end walls) forebay type. By comparison, Ensminger identified three classes and eighteen types of forebay barns.

### Amos Long Jr., *The Pennsylvania German Family Farm*

Published in 1972, *The Pennsylvania German Family Farm* is one of the great classics on Pennsylvania vernacular architecture. Henry Glassie praised it as one of the greatest treatises of its type ever written. The book contains twenty-six chapters on various building types found at German farms, including barns.

Long's discussion on the barn, based on mostly cursory visits, is primarily descriptive. No part of his book involved any ideas on evolution, but he rendered a good general description of the bank barn, log barn, frame barn, stone wall barn, and brick-end barn. Other topics include framework construction, roof construction, forebay, barn floor, wall openings, ramps, lower levels (in bank barns), appendages, painting and decorations, date boards and date-stones, and folklore.

Like others, Amos Long stated, "The bank barn is an American product and has no direct counterpart in European culture" (Long 1972, 358). Perhaps he relied on statements made by Henry Glassie, whom he quoted. Nevertheless, Long's work provided a stimulus for further research on the Pennsylvania barn.

### Eric Arthur and Dudley Witney, *The Barn: A Vanishing Landmark in North America*

This is one of the finest illustrated books ever published on barns. Arthur was a professor of architecture at the University of Toronto. Witney is an illustrator and

Left to right: Joseph Glass, John Heyl, and Robert Ensminger are pioneers in telling the story of barns in Pennsylvania.

photographer. Published in 1972, it is divided into four parts. Part one consists of illustrations of barns in a general sense to set the mood. Part two has five chapters, each one devoted to a particular barn type. Chapter three focuses on the Pennsylvania Barn with an oblique look at Ontario, which has a number of these barns. The authors posit that of all the bank style barns on the continent, the Pennsylvania barn is the acme of expression. Part Four looks at barn details.

The most illustrated barn, and likely their favorite, is the superb circa 1810 Dalziel double log crib forebay barn at Black Creek Pioneer Village near Toronto. Other barns are in Chester, Bucks, and Lancaster counties, among other places. They illustrate some other large stone barns. One error was their reliance on the statements of Thomas Wertenbaker about the area of Europe that gave rise to the forebay barn.

## Berenice Ball, *Barns of Chester County, Pennsylvania*

Berenice M. Ball's limited-edition 1974 book was a remarkable treatise in that it specialized in barns in only one area, the county just northwest of Philadelphia settled by the English. She covered many barns in the county, most of them of stone construction. Although the county is overflowing with all kinds of barns, she made no attempt at categorizing them. This was a woman who just loved barns and her book was a glorification of them. But she understood that they had structural integrity and devoted a chapter to framing. This book preserved a number of images of the area's barns and generated interest in the 1970s and beyond.

Dudley Witney of Guelph, Canada, has been a photographer for several decades and is still going strong at age ninety.

## Alan Keyser and William Stein, "The Pennsylvania German Tri-Level Ground Barn"

Until 1975, no treatise was ever written on the Ground barn. One of the earliest known barn forms in Pennsylvania, the German Ground barn was also called *grundscheier* or *boddamscheier* (bottom barn). The article by Keyser and Stein provided good insight into this generally diminutive barn. Some observers call their work a classic. Potentially confusing to the general reader is that these barns are actually one-level structures, but the early ones often have three sections built on different elevations. They were likely first built in the first quarter of the eighteenth century, and possibly before.

Robert Ensminger stands at the front of a farm lane near Jonestown in Lebanon County that leads to the first log barn he visited in the late 1970s.

## Jerry Apps and Allen Stang, *Barns of Wisconsin*

Ensminger wrote an article that came out in *Pennsylvania Folklife* in 1983, "A Comparative Study of Pennsylvania and Wisconsin Forebay Barns." This was a study that had never been attempted prior to his researches in the Midwest. Therefore, statements in the 1977 published book of Apps and Stang—*Barns of Wisconsin*—that came out six years before Ensminger's article would seem to have some particular relevance.

Apps and Stang state that "The traditional Pennsylvania bank barn, found in southern Wisconsin, and particularly in Green County, may be traced to Europe." But then they reiterate exactly what Wertenbaker said in his 1938 book about the highlands of Upper Bavaria and the other places he discussed. It is clear that Wertenbaker's strong statements, as erroneous as they were, were a source of several future scholars' incorrect assertions in their works.

## Stanley Schuler, *American Barns: In a Class by Themselves*

This 1984 book is replete with photographs of many types of barns all over the US. It serves readers who want a broad overview of barns across the nation, without in-depth analysis. The first chapter on bank barns includes about two dozen views of Pennsylvania forebay barns, many of them of stone.

## Terry Jordon, *American Log Buildings: An Old World Heritage*

Terry Jordan was a renowned author and a Walter Prescott Webb Professor in the Department of Geography at the University of Texas in Austin. He did extensive research in numerous European areas. Both he and Robert Ensminger in 1978 conducted independent field surveys in Switzerland. Jordan found many forebay barns in certain cantons in the Swiss mother country. Without any consultation, he came to nearly identical conclusions as Ensminger did regarding the origin of the Pennsylvania forebay barn.

Jordan's book, published in 1989, contains references to Pennsylvania barns and thoughts about their prototypes. The six-chapter book includes three chapters on the barn forms of specific areas in Europe. He says that certain features and elements of barns in each of the three areas are reminiscent of what he called Midland forms of log construction in America. He found barns in a few European areas that closely resembled the Pennsylvania barn in its log form. The form that was most similar was in Pratigau, a region of Graubunden in eastern Switzerland.

## Robert Ensminger, *The Pennsylvania Barn: Its Origin, Evolution, and Distribution in North America*

Robert Ensminger's book has provided by far the greatest contribution to Pennsylvania barn studies since the first edition appeared in 1992. No one will ever replace Eric Sloane as the pioneering author of North American barns; however, Ensminger's book is the first comprehensive tome by a North American barn author. The first edition was 200 hundred pages. In 2003, he produced the second edition that added a hundred pages of text and illustrations.

In 1975, he traveled to Europe to investigate possible origins of settlers in Europe who came to Pennsylvania in the eighteenth century. At the end of his trip, he concluded that the Rhineland-Palatinate area of Germany was the source of many Pennsylvania-German settlers but was devoid of forebay prototypes. He discovered, however, that many Ground barns existed there. Also of major significance was that a broad band of forebay bank barns could be seen in central and eastern Switzerland.

On another trip to Alpine Europe in 1978, he found that forebay barns occurred extensively beyond Pratigau but discontinuously into other cantons, and also into Austria in the Tyrol and in Slovenia in northern Yugoslavia. This sojourn, coupled with research on the literature, convinced him that the log forebay bank barns of Pratigau provided the best prototype of the earliest forebay bank barn in the Pennsylvania countryside.

Armed with extensive knowledge taken from many field trips in Pennsylvania and several other states in the United States and Canada, Ensminger wrote his five-chapter barn book. He discussed the origin, classification, evolution, diffusion, distribution, and future of the Pennsylvania barn. He has received wide acclaim for his book and thousands of readers and preservationists have been inspired by his work. His name has become almost synonymous with barns and barn preservation in Pennsylvania.

Ensminger focused his writings on the Pennsylvania or forebay barn. While a number of features are common to all the barns in the several types, the differences are distinctive enough that they demand a broad classification system, with the forebay barn as only one barn style, albeit an important one. His legacy is profound and will persist for generations to come.

## Marian Moffet and Lawrence Wodehouse, *East Tennessee Cantilever Barns*

This 1993 publication is one of the best scholarly books on a particular barn type in North America. The barns featured have significant cantilevered upper floor sections above lower-level log cribs, hence the name cantilever barn. These log barns come in five major floor plans: as a two-crib double cantilever, a two-crib cantilever, half-double cantilever, single-crib double cantilever, and a four-crib cantilever. All but the fourth type have a central runway or wagon passageway.

Several American barn types have been regarded as a possible prototype of the East Tennessee cantilever barns. This includes the forebay bank barns of the Pennsylvania Germans, among others such as log frontier forts or blockhouses that had cantilevers on all four sides. The authors acknowledge the possibility that the forebay barn influenced the cantilever barn, but they believe that the blockhouses may have provided specific inspiration for the region's first cantilever barns.

The Tennessee cantilever log barn may have had its origin in the Pennsylvania forebay barn. More research is needed to establish the possible connections between the two barn styles.

## Jerry A. Clouse, *A Study of Agricultural/Vernacular Architecture of Central and Southwestern Pennsylvania*

Although this book came out more than twenty years ago (1995), it remains the best book on regional variations of barns in Pennsylvania. It is replete with sketches of framing units found in the southwest part of the state. He discusses major and minor architectural features of the regional barns. A number of farm and building surveys are included. The 186-page book was not widely distributed but it is something of a classic.

# Bibliography


Abrams, Lawrence and Kathleen. *Salvaging Old Barns & Houses: Tear It Down and Save the Pieces*. New York: Sterling Publishing Company, Inc., 1983.

Adams, Harry C. *The Direct Tax of 1798 in the Second District Third Division of Pennsylvania: 12 Townships in the Upper Part of Bucks County*. Abstracted and compiled by Harry Adams. Bedminster, PA: Adams Apple Press, 1994.

Andrews, Francis B. *Medieval or "Tithe" Barns*. Birmingham and Midlands Archaeological Society, 1901.

Apps, Jerry, and Allen Strang. *Barns of Wisconsin*. Madison, WI: Tamarack Press, 1977.

Arthur, Eric, and Dudley Witney. *The Barn: A Vanishing Landmark in North America*. Toronto: M. F. Feheley Arts Co., 1972.

Babcock, Richard. *Barns of Roots America Disassembled, Researched, Restored, and Resurrected, by The Man Who Did It All*. Published by Richard Babcock, 1989.

Babcock, Richard, and Lauren R. Stevens. *Old Barns in the New World: Reconstructing History*. Lee, MA: Berkshire House Publishers, 1996.

Baillie, M. *Tree Ring Dating and Archeology*. Chicago: Chicago University Press, 1982.

Ball, Berenice. *Barns of Chester County Pennsylvania*. Kennett Square, PA: KNA, 1974.

Barcroft, Ambrose, ed. "Life in Bucks County in 1722/23." *Pennsylvania Magazine of History and Biography*, Vol. 27, 1960: 397–402.

Bechmann, Roland. *Trees and Man: The Forest in the Middle Ages*. Translated by Kathryn Dunham. New York: Paragon House, 1990: 175–77.

Black, Bryan, and Marc Abrams. "Analysis of Temporal Variation and Species-Site Relationships of Witness Tree Data in Southeast Pennsylvania." *Canadian Journal of Forest Research*. March 2001, 31, 3. Sciences Module.

Braun, E. Lucy. *Deciduous Forests of Eastern North America*. New York: Hafner Press, 1950.

Brunskill, Ronald. *Vernacular Architecture of the Lake Countries*. London: Faber and Faber, 1974.

Bryce, Emma. "The Witness Trees." *The New York Times*, September 18, 2012.

Bucher, Robert C. "Grain in the Attic." *Pennsylvania Folklife*, Vol. 13 No. 2, Winter 1962–1963.

Bucher, Robert C. "The Long Shingle." *Pennsylvania Folklife*, XVIII, No. 4, Summer 1969.

Bucher, Robert C., and Alan G. Keyser. "Thatching in Pennsylvania." *Der Reggeboge: Journal of the Pennsylvania German Society*, 1982.

Callahan, William, and Ed Cook. "Dendro-chronological Analysis of the Daniel B. Charpentier House, South 4th Street." Philadelphia, PA: April 2010.

Charles, F. W. B., with Mary Charles. *Conservation of Timber Buildings*. Oxford and Northampton: Alden Press Limited, 1984.

Chauncy, Charles. 1649. *New England Historical and Genealogical Register*. Vol. X, April 1856, No. 2: 118.

Chester County, Pennsylvania. Probate Papers, Nos. 144 and 203.

Clouse, Jerry A. *A Study of Agricultural/Vernacular Architecture of Central and Southwestern Pennsylvania*. Harrisburg, PA: Pennsylvania Historical and Museum Commission, 1995.

Cochecton, New York website, "The Lord High Admiral (And the Elephant Who Tied Up River Traffic)," Delaware River Basin Commission.

Cogbill, C. V. "Vegetation of the Pre-Settlement Forests of Northern New England and New York." *Rhodora* 2000, 102: 250–276

Constantine, Albert, Jr. *Know Your Woods*. New York: Charles Scribner's Sons, 1959.

Cook, Anne H., and Ann L. Snider. "If Only the Walls Could Talk: The Story of the Federal Barn." *Pennsylvania Heritage*. Vol. IX, No. 4, Fall 1983.

Cook, Ed. Email correspondence. February 24, 2016.

Courtenay, Lynn. "From the Aisled Hall to the Un-aisled Hall." Online - Early Roofing Systems in Northern Europe.

Cronon, William. *Changes in the Land: Indians, Colonists, and the Ecology of New England*. New York: Hill and Wang, 1983.

Cuff, D. J., editor et. al. *The Atlas of Pennsylvania* in Economic Geography. Clark University, Vol. 67, No. 2, April 1991.

Domesday Book of 1086. Manuscript held at The National Archives at Kew, London. Open Domesday site has manuscript available online.

Donmoyer, Patrick. *Hex Signs: Myth and Meaning in Pennsylvania Dutch Barn Stars*. Kutztown, PA: Pennsylvania German Cultural Heritage Center, 2013.

Dornbusch, Charles H., and John K. Heyl. *Pennsylvania German Barns*. Allentown, PA: Schlecter's: The Pennsylvania German Folklore Society, Vol. 21, 1956. 1958.

Endersby, Elric, Alexander Greenwood, and David Larkin. *Barn: The Art of a Working Building*. Boston and New York: Houghton Mifflin Company, 1992. 1: 2, 36.

Engel, Katherine Carte. *Religion and Profit: Moravians in Early America*. Philadelphia: University of Pennsylvania Press, 2009.

Ensminger, Robert F. "A Comparative Study of Pennsylvania and Wisconsin Forebay Barn." *Pennsylvania Folklife*, 32, 1983. No. 3: 256–270.

Ensminger, Robert F. *The Pennsylvania Barn: Its Origins, Evolution, and Distribution in North America*. Baltimore and London: The Johns Hopkins University Press, 1992.

Fitchen, John. *The New World Dutch Barn: A Study of Its Characteristics, Its Structural Systems, and Its Probable Erectional Procedures*. Syracuse: Syracuse University Press, 1968.

Fitchen, John, and Gregory D. Huber. *The New World Dutch Barn: The Evolution, Forms, and Structure of a Disappearing Icon*. Syracuse: Syracuse University Press, 2001.

Fletcher, Stevenson Whitcomb. *Pennsylvania Agriculture and Country Life 1640–1840*. Harrisburg, PA: Pennsylvania Historical and Museum Commission, 1971.

Fox, William F. *History of the Lumber Industry in the State of New York*. Harrison, NY: Harbor Hill Books, 1976.

Friedrich, Paul. *Proto Indo-European Trees*. Chicago: University of Chicago Press, 1970.

Futhey, J. Smith, and Gilbert Cope. *History of Chester County, Pennsylvania*. Philadelphia: Press of J. B. Lippincott and Co.: Louis H. Everts, 1881.

Gallagher, Sarah A. "1703–1903: Early History of Lambertville, New Jersey." Trenton, NJ: MacCrellish & Quigley, Printers.

Gehring, Charles. New Netherland Institute in Albany, New York. Conversation on April 12, 2016, about the earliest-referenced Dutch-American barns in North America and Director Kieft's citation to a house-barn of the 1630s.

Germantown Historical Society has image of Unruh Homestead. Catalogue number: 2012.6 1059.

Glass, Joseph W. "The Pennsylvania Culture Region: A Geographical Interpretation of Barns and Farmhouses." PhD thesis, Department of Geography, State College, Pennsylvania State University, 1971.

Glass, Joseph W. *The Pennsylvania Culture Region: A View from the Barn*. Ann Arbor, MI: UMI Research Press, 1986.

Glassie, Henry. "The Double Crib Barn in South-Central Pennsylvania" Part 4. *Pioneer America* 2, No. 2, July 23–24, 1970.

Glassie, Henry. "The Pennsylvania Barn in the South" Part 1. *Pennsylvania Folklife* 15, No. 2, 1965: 8–19.

Glassie, Henry. "The Pennsylvania Barn in the South" Part 2. *Pennsylvania Folklife* 15, No. 4, 1966: 12–25.

Glassie, Henry. "The Variation of Concepts within Tradition: Barn Building in Otsego County, New York" *Geoscience and Man*. Vol. 5, Man and Cultural Heritage edited by Bob F. Perkins, 177–235. Baton Rouge: Louisiana State University, 1974.

Harlow, William, PhD, and Ellwood S. Harrar, PhD. *Textbook of Dendrology: Covering the Most Important Forest Trees of the United States and Canada*. New York: McGraw-Hill Book Company, 1969.

*History of the Church of the Brethren of the Eastern District of Pennsylvania:* 1915 Committee Appointed by District Conference. Lancaster, PA: New Era Printing Co.

Horn, James. *A Land As God Made It: Jamestown and the Birth of America*. New York: Basic Books, 2005.

Huber, Gregory D. "Barn Roof Timbers that Helped Build Early German-American Culture in Pennsylvania: *Liegender Dachstahl* and other Roof Supports." *Der Reggeboge - Pennsylvania German Society*. Vol. 42, No. 1 issue, 2008: 23–52.

Huber, Gregory D., foreword and editor. *Barns: A Close-Up Look*. East Petersburg, PA: Fox Chapel Publishing, 2011.

Huber, Gregory D. "Barns in Lehigh County in 1798." *The Forebay Post:* The Historic Barn and Farm Foundation of Pennsylvania, Fall 2012.

Huber, Gregory D. "Barns in Macungie Township: Eighteenth Century Compared to the Present Day." Lower Macungie Township Historical Society, 2013.

Huber, Gregory D. "Behind the Threshing Doors: An Inside Look at Some of the Earliest Barns in Pennsylvania." *Material Culture*: Vol. 36, No. 2, Fall 2004: 22–55.

Huber, Gregory D. "Dendrochronology." *Heller Homestead News*. Vol. 16, No. 2, Summer 2009.

Huber, Gregory D. "Dutch Anchor-beams in Pennsylvania Barns." *Dutch Barn Preservation Society Newsletter*. Vol. 6, No. 1, Spring 1993.

Huber, Gregory D. *Dutch Barn Research Journal*. Vol. 1 and 2, 1992: 1–111 and Vol. 3 and 4, 1995: 1–80.

Huber, Gregory D. "Extended Forebay Barns with Front Straw Sheds in Longswamp." Special edition of *Longswamp Township Historical Society Newsletter*, December 2014.

Huber, Gregory D. "Frame Barns in the Eighteenth Century." *The Forebay Post:* The Historic Barn and Farm Foundation of Pennsylvania, Spring 2013.

Huber, Gregory D. "Framing Techniques as Clues to Dating in Certain Pre-Revolutionary Dutch Barns: Major and Minor Rafter Systems, Lapped Dove-tail Joinery, *Verdiepinghs* and other Traits." *Material Culture*: Vol. 29, No. 2, Summer 1997.

Huber, Gregory D. "From the Distant Past: Early Type *Fachwerk* German Ground barn in Central New Jersey." *Der Reggeboge - Pennsylvania German Society*. Vol. 42, No. 1 issue, 2008.

Huber, Gregory D. "Ground Barns That Hide and Puzzle Us in Longswamp." *Longswamp Township Historical Society Newsletter*, April/May 2015.

Huber, Gregory D. "Ninety Degree Roof Rotations in New Jersey Dutch Barns." *Material Culture:* Vol. 31, No. 1, Spring 1999: 1–20.

Huber, Gregory D. "Oddities Observed: The Old Ground Barns of Oysterdale." *Historical Review of Berks County*, Winter 2010–2011.

Huber, Gregory D. "One-level Ground barns in Longswamp Township." *Longswamp Township Historical Society Newsletter*, April 2014.

Huber, Gregory D. "Piedmont Lowlands of Lancaster County: Is This the Birthplace of the Iconic Schweitzer Barn Style?" Historic Barn and Farm Foundation of Pennsylvania Barn Tour Guide Booklet, June 2015.

Huber, Gregory D. "The Swing Beam Barn in the New World Dutch Cultural Hearth." *Dutch Barn Preservation Society Newsletter*, Spring 2011, Vol. 24 Issue 1.

Huber, Gregory D., Margaret Bye Richie, and John D. Milner. *Stone Houses: Traditional Homes of Pennsylvania's Bucks County and Brandywine Valley*. New York: Rizzoli International Publications, Inc., 2005.

Hubka, Thomas C. *Big House, Little House, Back House, Barn: The Connected Farm Buildings of New England.* Hanover and London: University Press of New England, 1984.

Innocent, C. F. *The Development of English Building Construction.* David and Charles Reprints, 1971.

Jicha, Hubert F., III. "Bank Barns in Mill Creek Hundred, Delaware." Honor's Thesis at the University of Delaware, 1984.

Jordan, Terry G. *American Log Buildings: An Old World Heritage.* Chapel Hill, NC: University of North Carolina Press, 1985.

Jordan, Terry G., and Matti Kaups. *The American Backwoods Frontier: An Ethnic and Ecological Interpretation.* Baltimore and London: The Johns Hopkins University Press, 1989.

Keever, Catherine. "Distribution of Major Forest Species in Southeastern Pennsylvania." *Ecological Monographs*, Summer 1973.

Kershner, Bruce, and Robert T. Leverett. *The Sierra Club Guide to the Ancient Forests of the Northeast.* San Francisco: Sierra Club Books, 2004.

Keyser, Alan, and William Stein. "The Pennsylvania German Tri-Level Ground Barn." *Der Reggeboge: Quarterly of the Pennsylvania Society.* Vol. 9, No. 3, 4: 1–25.

Klamkin, Charles. *Barns: Their History, Preservation, and Restoration.* New York: Hawthorn Books, Inc., 1973.

Lauver, Joseph E., III. "The Pennsylvania Log Barn in the Mahantango Valley." Bristol, RI: Roger Williams University, 1993.

Learned, Marion. "The German Barn in America." University of Pennsylvania Lectures Delivered by Members of the Faculty in the Free Public Lecture Course, 338–49. Philadelphia: University of Pennsylvania, 1915.

LeVan, Kenneth. "Building Construction and Materials of the Pennsylvania Germans." Harrisburg, PA: Vernacular Architecture Forum Meeting, 2004.

Leydet, Francois (text), and James Rose (photographs and others). *The Last Redwoods and the Parkland of Redwood Creek.* Sierra Club: Ballantine Books, 1969.

Logan, William Bryant. *Oak: The Frame of Civilization.* New York and London: W.W. Norton and Company, 2005.

Long, Amos. *The Pennsylvania German Family Farm.* Breinigsville, PA: Pennsylvania German Society, Vol. 6, 1972.

Lord, Arthur. "Barns of Lancaster County: 1798." *Journal of the Lancaster County Historical Society*, Vol. 77, No. 1, Hilarymas, 1973.

Lorimer, C. G. "The pre-settlement forest and natural disturbance cycle of northeastern Maine." *Ecology* 58, 1977: 139–148.

Lower Macungie Historical Society. *A History of Lower Macungie Historical Society, Lehigh County, Pennsylvania*: Second Edition. Wescosville, PA: Always Printing, 1996.

Malone, Joseph J. *Pine Trees and Politics: The Naval Stores and Forest Policy in Colonial New England 1691– 1775.* Seattle, WA: University of Washington Press, 1964.

Maloof, Joan. *Among the Ancients: Adventures in the Eastern Old-Growth Forests.* Washington, DC: Ruka Press, 2011.

Marshall, Jeffrey, and Willis M. Rivinus. *Barns of Bucks County.* Endorsed by Heritage Conservancy & The Bucks County Audubon Society, 2007.

Miles, Daniel. "Refinements in the Interpretation of Tree-Ring Dates for Oak Building Timbers in England and Wales." *Vernacular Architecture*, Vol. 37, 2006: 84–96.

Mills, Edward. "A Brief History of Coppicing." www.smallwoods.org.uk.

Moffet, Marian, and Lawrence Wodehouse. *East Tennessee Cantilever Barns.* University of Tennessee Press, 1993.

Mosseler, A., I. Thompson, and B. Pendrel. "Old Growth Forests in Canada: A Science Perspective." Presentation at a symposium in Canada, 2001.

Neccai, Terry. *The Pennsylvania Farm Landscape as a Historic Resource: The Forces That Gave It Shape*: Prepared under the auspices of the Athenaeum of Pennsylvania for the 2011 Charles I. Peterson Fellowship: Philadelphia and Monongahela, January 2012, Chapter 3: 120.

Nixon, Kevin C. "The Oak (*Quercus*) Biodiversity of California and Adjacent Regions." USDA Forest Service Gen. Tech. Rep. PSW-GTR-184, 2002.

Nutting, Wallace. *Pennsylvania Beautiful.* Framingham, MA: Old America Company, 1924.

Perkins, Don. *The Barns of Maine.* Charlestown, SC: The History Press, Vol. 15, 2012.

Pliny the Elder. *The Natural History.* John Bostock, MD, FRS, H. T. Riley, Esq., BA, Ed.

Podvia, Mark. "The Honorable Frederick Watts: Carlisle's Agricultural Reformer." *17 Penn State Environmental Law Review* 299, 2009.

Porteous, Alexander. *The Forest in Folklore and Mythology.* Mineola, New York: Dover Publications, Inc., 2002.

Preston, Richard. *The Wild Trees: A Story of Passion and Daring.* New York: Random House, 2007.

Rackham, Oliver. *The Last Forest: The Story of Hatfield Forest.* London: M Dent & Sons Ltd., 1989.

Rackham, Oliver. *Woodlands.* London: Collins, 2006.

Roan, Ab. Personal conversation; 2002.

Rooke, Hayman. *Descriptions and Sketches of Some Remarkable Oaks in the Park at Welbeck in the County of Nottingham.* London: Printed by J. Nichols, 1790. Transcribed in 2010 by John Palmer.

Russell, Emily W. B. "Indian-set Fires in the Forests of Northeastern United States." *Ecology.* Vol. 64, 1983: 78–88.

Schiffer, Margaret Berwind. *Survey of Chester County, Pennsylvania, Architecture: 17th, 18th, and 19th Centuries.* Atglen,PA: Schiffer Publishing, 1984.

Schmithüsen, F., and W. Zimmermann. "Forestry Case Study Switzerland." Pelkonen et al. (eds.) Forestry in Changing Societies in Europe. 1999.

Schuler, Stanley. *American Barns: In a Class by Themselves. Atglen, PA:* Schiffer Publishing, 1997.

Shaw, Edward. *Civil Architecture.* Cleveland, OH: Jewett, Proctor, and Worthington, 1852.

Shoemaker, Alfred. *Hex No!* Lancaster: Pennsylvania Dutch Folklore Center, Franklin and Marshall College, 1953.

Shoemaker, Alfred. *The Pennsylvania Barn.* Lancaster: Pennsylvania Dutch Folklore Center, Franklin and Marshall College, 1955.

Shurtleff, Harold. *The Log Cabin Myth: A Study of the Early Dwellings of the English Colonists in North America.* Cambridge, MA: Harvard University Press, Vol. 126, 1939.

Sloane, Eric. *A Museum of Early American Tools.* New York: Funk and Wagnalls, 1964.

Sloane, Eric. *A Reverence for Wood.* New York: Funk and Wagnalls, 1965.

Sloane, Eric. *American Barns and Covered Bridges.* New York: Funk and Wagnalls, 1954.

Sloane, Eric. *An Age of Barns.* New York: Funk and Wagnalls, 1966.

Sloane, Eric. *Eighty: An American Souvenir.* New York: Dodd and Mead, 1985.

Sloane, Eric. *I Remember America.* New York: Funk and Wagnalls, 1971.

Tudge, Colin. *The Tree: A Natural History of What Trees Are, How They Live, and Why They Matter.* New York: Three Rivers Press, 2005.

van Ravenswaay, Charles. *America's Age of Wood.* Worcester, Massachusetts: American Antiquarian Society, 1970.

Vaughan, Alden. *American Genesis: Captain John Smith and the Founding of Virginia.* Boston: Little Brown and Co., 1975.

Visit to New England in April 2013 to David E. Lenoue Inc. Building and Design, Stockbridge, MA, to see the 1690s remnant English barn.

Visser, Thomas Durant. *Field Guide to New England Barns and Farm Buildings.* Hanover and London: University Press of New England, 1997.

von Simson, Otto. *The Gothic Cathedral: Origins of Gothic Architecture and the Medieval Concept of Order.* New York and Evanston: Harper and Row, 1956.

Watts, Frederick. "The Pennsylvania Barn": cited in York County Star and Wrightsville Advertiser, from *Agricultural Report* 1864: 289–298, illus.

Website: The Industrial Revolution: Transportation Revolution (http://firstindustrialrevolution.weebly.com.

Wenger, Samuel. "1710 Pequea Settlement Tour Resource Information Booklet." Mennonite Historical Society, 1997.

Wertenbaker, Thomas. *Patrician and Plebeian in Virginia.* Charlottesville, VA: The Michie Printers, 1910.

Wertenbaker, Thomas. *The Founding of the American Civilization the Middle Colonies.* New York and London: Charles Scribner's Sons, 1938.

Wertenbaker, Thomas. *The Planters of Colonial Virginia.* Princeton: Princeton University Press, 1922.

Weslager, C. A. *The Log Cabin in America: From Pioneer Days to the Present.* New Brunswick, NJ: Rutgers University Press, 1969.

Whitaker, Tim. "The Bank Barns of Cumbria: an Overview." *Journal of the Historic Farm Buildings Group*, Vol. 15, 2001.

Whitaker, Tim. *The Origin and Development of the Bank Barn in Cumbria.* Manchester, England: Master of Philosophy Thesis, University of Manchester, 1989.

Williams, Michael. *Americans and Their Forests: A Historical Geography.* Cambridge: Cambridge University Press, 1989.

Willis, Reverend Robert. *A Description of the Sextry Barn at Ely, Lately Demolished.* Cambridge, England: Deighton, 1843.

Wills, Charles A. *A Historical Album of Pennsylvania.* Millbrook Press, 1996.

Yoder, Don, and Thomas E. Graves. *Hex Signs: Pennsylvania Dutch Barn Symbols and Their Meaning.* New York: E. P. Dutton, 1989.

Zon, Raphael. *The Vanishing Heritage* (unpublished) in the research compilation file: Records of the Forest Service: Record Group 95 at the National Archives.

# Index

## V



## W